Polymeric Materials for
Electronic Packaging

IEEE Press
445 Hoes Lane
Piscataway, NJ 08854

Polymeric Materials for Electronic Packaging

Shozo Nakamura
Doctor of Engineering, Professor Emeritus
Hiroshima Institute of Technology, and Corporate Technical Adviser, Japan

IEEE PRESS
WILEY

Published by John Wiley & Sons, Inc., Hoboken, New Jersey.
Published simultaneously in Canada.

Library of Congress Cataloging-in-Publication Data
Names: Nakamura, Shozo (Professor Emeritus), author.
Title: Polymeric materials for electronic packaging / Shozo Nakamura.
Description: Hoboken, New Jersey : Wiley, [2023] | Includes bibliographical references and index.
Identifiers: LCCN 2023013757 (print) | LCCN 2023013758 (ebook) | ISBN 9781394188796 (hardback) | ISBN 9781394188802 (adobe pdf) | ISBN 9781394188819 (epub)
Subjects: LCSH: Conducting polymers. | Electronic packaging.
Classification: LCC QD382.C66 N35 2024 (print) | LCC QD382.C66 (ebook) | DDC 620.1/9204297–dc23/eng20230708
LC record available at https://lccn.loc.gov/2023013757
LC ebook record available at https://lccn.loc.gov/2023013758

Cover Design: Wiley
Cover Image: © Nikola Ilic/Getty Images

Set in 9.5/12.5pt STIXTwoText by Straive, Pondicherry, India

Contents

About the Author *ix*
Preface *xi*

1 Basics of Semiconductor *1*
1.1 Development of Semiconductors *1*
1.2 Analysis of Semiconductors Materials *5*

2 Basics of Polymer Materials *9*
2.1 Polymer Material *9*
2.2 Types and Classification of Polymer Materials *10*
2.3 General Properties of Polymer Materials *12*
2.4 Summary *13*

3 Basics of Elastic Theory *15*
3.1 Elasticity *15*
3.2 Stress and Strain *15*
3.3 Finite Element Method Analysis (FEM Analysis) *16*
3.4 Governing Equation of Elastic Body *18*
3.5 Law of Elastic Breakage *19*
3.6 Plane Stress and Plane Strain *21*

4 Stress Evaluations with Defects *23*
4.1 Difference from Strength of Materials *23*
4.2 Stress Concentration and Stress Intensity Factor *24*

5 Basics of Viscoelasticity *27*
5.1 About Viscoelasticity *27*
5.2 Elasticity, Viscosity, and Viscoelasticity *28*
5.3 Stress and Strain Response *29*

5.4 Mechanical Model Representing Viscoelastic Properties *32*
5.5 Conceptual Formula for Creep and Stress Relaxation *35*
5.6 Master Curve and Time–Temperature Conversion Rule *38*
5.7 Approximation of Master Curve *39*
5.8 Superposition Principle and Basic Equations *40*
5.9 Simple Model of Generating Thermal Stress and Strain *41*

6 Measurement of Viscoelastic Properties *43*
6.1 Dynamic Viscoelasticity *43*
6.2 Measurement Method *43*
6.3 Complex Modulus and Mechanical Model *44*
6.4 Dispersion and Absorption by Frequency *46*
6.5 Actual Measurement Example *48*

7 Design Issues of LSI Packages *49*
7.1 Introduction *49*
7.2 Trends and Issues of LSI Packages *51*

8 Validity of Viscoelastic Analysis *55*
8.1 Introduction *55*
8.2 Structure of Laminated Body *56*
8.3 Analysis Method *61*
8.4 Cooling Experiment of Laminated Body *61*
8.5 Analysis Results and Experimental Values *63*
8.6 Conclusion *67*

9 Application to CSP-μBGA *69*
9.1 Introduction *69*
9.2 Structure and Modeling of CSP-μBGA *69*
9.3 Material Property Values Used for Analysis *70*
9.4 Material and Structure Optimization Design by VESAP Analysis *73*

10 Thermal Stress and Warpage Behavior During Cooling Process of Three-Layer Laminate *79*
10.1 Introduction *79*
10.2 Structure of LSI Package *79*
10.3 Three-Layer Viscoelastic Laminate Model *80*
10.4 Elucidation of Warpage Deformation Behavior by VESAP Analysis *81*

11 Warp Deformation Behavior From Heating to Cooling *91*
11.1 Introduction *91*
11.2 Two-Layer Laminate With Epoxy Resin/FR-4 Substrate *92*
11.3 Two-Layer Laminate with Epoxy Resin/Steel *101*
11.4 Three-Layer Laminate with Steel/Epoxy Resin/Printed Board *104*
11.5 Analysis Experiment of Four-Layer Laminate *108*

12 Deformation Prediction Method Considering Curing Shrinkage of Resin *119*
12.1 Introduction *119*
12.2 Examination the Procedure and the Way of Thinking *120*
12.3 Contents of VESAP Analysis *122*
12.4 Simple Prediction Formula for Calculating Curing Warpage *123*
12.5 Warp Deformation Experiment *128*
12.6 Theoretical Prediction of Warpage Deformation due to Hardening and Heat *130*

13 Changes in Material Properties and Deformation Behavior Due to Thermal Degradation *133*
13.1 Purpose and Background *133*
13.2 Experimental Case I *134*
13.3 Experiment of Case II *139*

14 Simple Evaluation Method for Deformation of Viscoelastic Body *147*
14.1 Introduction *147*
14.2 Derivation of Simple Formula *147*
14.3 Practical Method *151*
14.4 Determining the Curing Temperature of the Resin *152*
14.5 Effect of Epoxy Resin Thickness on Heat Generation Temperature *153*
14.6 Conclusion *156*

15 Effect of Cooling Rate on Warpage Behavior of Laminates *159*
15.1 Warp Deformation Experiment *159*
15.2 VESAP Analysis *160*
15.3 Results and Considerations *163*
15.4 Final Warp Deformation and Residual Warp Deformation *166*
15.5 Estimation Mechanism Between Cooling Rate and Deformation of Laminate *168*
15.6 Conclusion *169*

Appendix A Development of Viscoelastic Analysis Software (VESAP) *171*
A.1 Development Needs and Concepts *171*
A.2 Derivation of Basic Formula for Analysis *172*
A.3 Contents of the Developed VESAP Software *175*

Bibliography *179*
Index *185*

About the Author

Shozo Nakamura, Professor Emeritus, Hiroshima Institute of Technology, and Corporate Technical Adviser, Japan. Professor Nakamura graduated from the Department of Mechanical Engineering, Hiroshima Institute of Technology in 1971 after which he joined Hitachi Limited. He became a Doctor of Engineering in 1989, and a Professor in 2000 at the Department of Intelligent Mechanical Engineering, Hiroshima Institute of Technology. He established the Nakamura Technical Research Institute in 2019. He was given the title of Professor Emeritus in 2018. He has written numerous publications on viscoelasticity.

Preface

After graduating from university, I joined Hitachi, Ltd. and engaged in research on the viscoelastic analysis of the strength of general-purpose resins, stress, and deformation of semiconductor encapsulation resins at the Yokohama Research Institute (later, Production Technology Research Institute). After that, moving from Hitachi to university, I was involved in research on viscoelasticity analysis for half a century throughout Hitachi and college.

Currently, smartphones, iPads, notebook PCs, etc., are widespread. Many semiconductor packages are frequently used in these electronic products, and resins are used in these semiconductor packages without exception.

The viscoelastic properties of this resin greatly affect the strength and deformation reliability of semiconductor packages, and optimal design technology based on resin viscoelastic analysis is indispensable for improving the quality of electronic products that make heavy use of semiconductor packages.

I was troubled by the viscoelastic properties of the resin when evaluating the reliability of semiconductor packages in the past. That is, as described in detail in the text, the semiconductor package is composed of many different materials such as LSI chips, resins, metals, solders, and substrates, and in particular, resins play various roles such as sealing LSI chips and buffering thermal stress.

In the development of semiconductor packages that are becoming lighter, thinner, shorter, and smaller, it is a major technical issue to prevent minute cracks and peeling at the resin and it adhesive interface, and to prevent warpage deformation due to the heat load during manufacturing. The optimal design of the materials, structures, and thermal processes that make up the semiconductor package was indispensable.

However, the stress simulation analysis half a century ago is based on the elastic theory, and it is a fatal problem that the viscoelastic behavior in which the elastic modulus and linear expansion coefficient of the resin are greatly affected by time and temperature is not taken into consideration.

As an episode under development at Hitachi, the warp direction of the actual product was the exact opposite of the analysis result, and I clarified that the cause of the defect is largely related to the viscoelastic property of the resin.

From this bitter experience, I strongly felt the need to develop a viscoelastic analysis and persuaded my boss to start the development. As a result, after 10 years, I developed viscoelastic analysis software and named it VESAP (Visco Elastic Stress Analysis Program). This VESAP is useful for the rapid and optimized design of materials, structures, and thermal processes to prevent defects in strength and deformation of electronic components including all resins such as semiconductor packages.

My aim in VESAP development was to focus on analysis software that can be easily and quickly utilized by nonspecialized researchers and engineers, and that contributes to shortening the product development period as an immediate force. Of course, the basis of the idea of analysis software is supported by the theory of viscoelasticity.

After that, I received technical consultations from many major companies in the electronics field such as semiconductor packages, electronic components, and resin substrates. The contents are concrete, such as

1) what kind of physical properties are optimal for preventing cracks and warp deformation at the joint interface, and
2) what physical properties are involved in strength and warp deformation. It was a quantitative requirement.

By the way, the connection between viscoelasticity and me is the encounter with Professor Yasushi Miyano, who was transferred from Hitachi, Ltd. to a professor at Kanazawa Institute of Technology, and Professor Takeshi Kunio, an emeritus professor at Keio University.

While working at Hitachi from early morning to late night, I desperately wrote research papers and presented at academic conferences without returning home, and obtained a doctorate in engineering. This is also a result of the guidance of both professors. Both the professors were strict but passionate guides to research and warm personalities, full of humanity. I would like to express my deep gratitude to both professors.

On the other hand, my relationship with semiconductor packages was the encounter with Dr. Gen Murakami, who was the development leader at the Musashi Work of Hitachi, Ltd. As described in Chapter 1, the semiconductor package has changed from a pin insertion type to a surface mount type represented by LOC (Lead On Chip) and QFP (Quad Flat Package), and is a smaller and thinner CSP (Chip Scale Package). It has greatly developed and changed into a type of BGA (Ball Grid Array)-CSP. Dr. Murakami is a pioneer of this semiconductor package.

I thank Dr. Murakami for his guidance owing to which I was able to contribute to the development and application of VESAP, and I would like to express my sincere gratitude to him.

In addition, although VESAP was developed during the Hitachi era, many issues such as its application limits could be clarified through basic research with university students. Thanks to these students.

Furthermore, I would like to express my sincere gratitude to Sandra Grayson – Senior Commissioning Editor, Becky Cowan – Senior Editorial Assistant, and Jayashree S – Managing Editor, and Mustaq Ahamed – Content Refinement Specialist, for their kind and courteous guidance and advice on the series of tasks from application to contract in publishing this book, and Wiley for their guidance in publishing this book.

Finally, if this book is of any help to young engineers and researchers in the electronics field such as semiconductor packages containing resin around the world, it would be an unexpected joy to me.

12 May 2023 *Shozo Nakamura, Japan*

1

Basics of Semiconductor

1.1 Development of Semiconductors

Various semiconductor packages and electronic devices are used in various electronic devices such as smartphones, mobile phones, tablets, iPads, and notebook PCs these days.

In recent years, in the development of these electronic devices, the development of new materials and manufacturing methods as well as the optimization of their mounting technologies are wide ranging, and the competitiveness of product development among companies is extremely fierce. The basic performance of semiconductor packages, which are often used in electronic devices, is the function of efficiently connecting large-scale integrated (LSI) circuit chips and printed circuit boards and the function of protecting LSI chips under usage in various environmental conditions. These are directly linked to the functions of electronic devices and LSIs, and therefore optimization is required in the selection of various materials that form semiconductor packages and the selection of highly reliable manufacturing conditions.

At the same time, optimal design of semiconductor package shapes and standards with an eye on global standardization is considered to be a key technology that will lead to the development of the electronic industry in the future. Against this background, the mounting form of semiconductor packages has also undergone changes as shown in Fig. 1.1a, b.

That is, as shown in Fig. 1.1a, b, in the 1970s, the semiconductor package was a pin insertion mounting type in which pins were inserted into holes provided in advance on the printed circuit board and solder-bonded, but in the 1980s, the surface of the printed circuit board was aimed at thinning. In the 1990s, it changed to a surface mount type that solders to a printed circuit board, and in the 1990s,

Polymeric Materials for Electronic Packaging, First Edition. Shozo Nakamura.

Published 2023 by John Wiley & Sons, Inc.

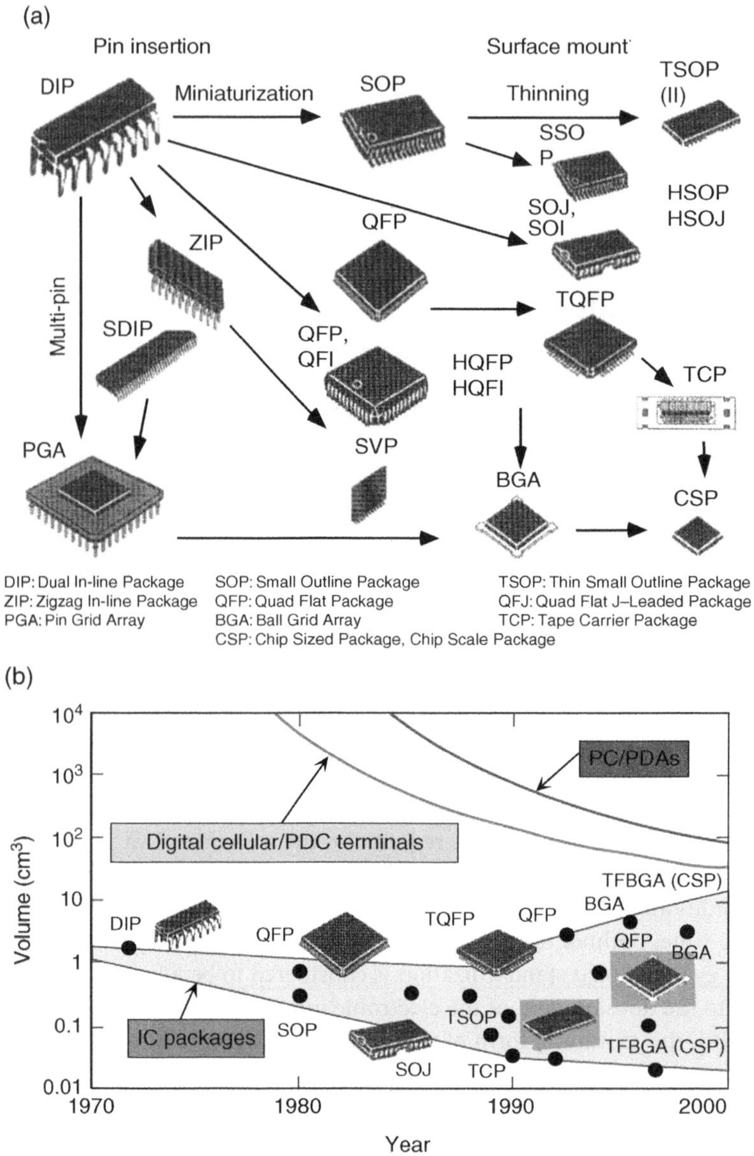

Figure 1.1 (a) Outline transition of LSI package. (b) Changes in volume of LSI packages.

it became an area array mount type so that a large number of semiconductor packages can be mounted on a printed circuit board (high-density mounting is possible).

Here, as examples of semiconductor packages representing various mounting types, there are dual in-line package (DIP) and pin grid array (PGA) for the pin insertion type, quad flat package (QFP), thin small outline package (TSOP), small outline J-leaded package (SOJ) for the surface mounting type, and ball grid array-chip scale package (BGA-CSP) for the area array mounting type.

At the same time, the shape and structure of the LSI package have undergone remarkable changes as shown in Fig. 1.1a, b and the structure of the CSP that realizes the ultimate lightness, thinness, shortness, and miniaturization.

In other words, the initial package form has changed from the pin insertion type represented by DIP to the surface mount type represented by QFP and has been further reduced in size and thickness to the utmost limit. It has been transformed into a CSP package.

In addition, the package thickness has been developed into a BGA type CSP package in which a large number of solder balls are placed on the bottom of the printed circuit board via a thin TSOP with a mounting height of about 1 mm for applications such as memory cards.

As a result, various semiconductor packages having such a CSP structure are widely used in current information terminal devices such as smartphones, mobile phones, tablets, and notebook computers.

The size of the LSI chip used for these also differs depending on the product, but in general, the area is small, such as about 15 mm × 30 mm to 5 mm × 10 mm, and the thickness is also thin, about 0.4–0.05 mm.

Old semiconductor packages (DIP, QFP, etc.) are generally composed of many dissimilar materials with different properties such as LSI chips, metal lead frames, solder, underfills, and sealing resins.

However, LSI packages with a CSP structure are often used in recent electronic devices, and an example of this is schematically shown in Figs. 1.2 and 1.3 with cross section of various LSI packages structure.

Each semiconductor manufacturer has developed various LSI packages, and it can be seen that each has a unique structure. It is sectional drawing of the mounting body of the multi-pin and lightweight structure used. Elastomer resin and encapsulant are used to protect the LSI chip from external forces, prevent cracks caused by thermal stress, and prevent warpage deformation of the semiconductor package itself.

Here, as a representative example, an example of five types of LSI packages is shown.

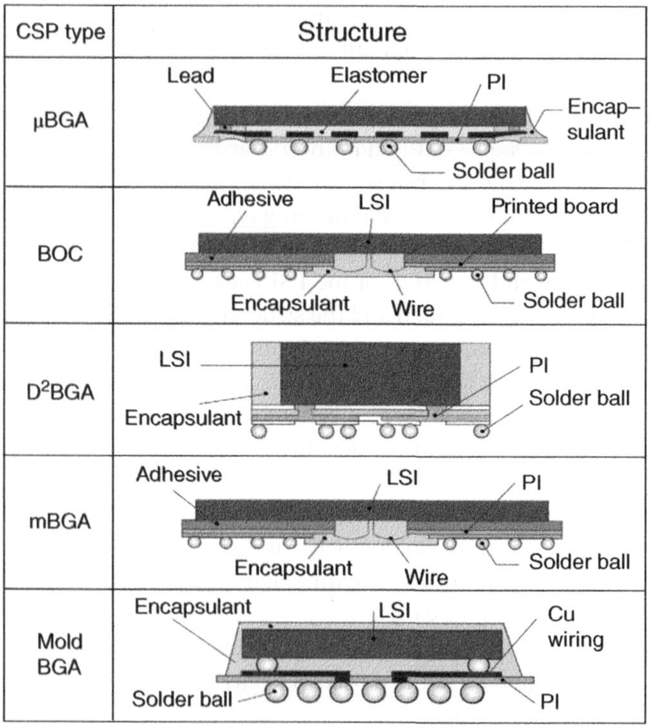

Figure 1.2 Examples of cross-sectional structures of various LSI packages.

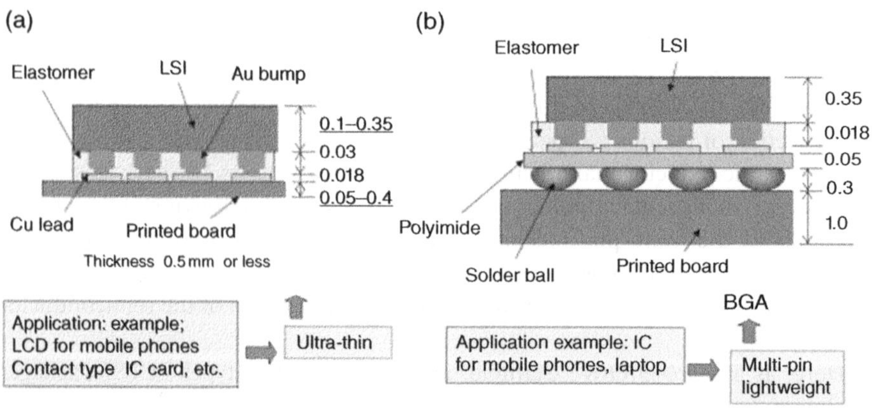

Figure 1.3 Cross-section example of LSI package. (a) FCA (flip chip attach). (b) TOP (tape on chip package).

1.2 Analysis of Semiconductors Materials

Figure 1.4 shows the cross-sectional structure of μBGA as an example. As shown in Fig. 1.4, μBGA has many different properties.

In general, sealing resins, elastomer, and underfill resins are materials having so-called viscoelastic properties whose thermal and mechanical properties change greatly with time and temperature, unlike metals and ceramics.

This viscoelastic behavior of various materials as shown in Fig. 1.3 greatly affects the performance and reliability of electronic devices that are finally manufactured and marketed. It can be said that the mechanical behavior of the resin affects these performance, quality, and reliability.

Therefore, it is indispensable for high-quality and highly reliable electronic equipment to properly understand the viscoelastic behavior of the resin to be used.

In LSI packages as shown in Figs. 1.2–1.4, an LSI chip made from a silicon wafer is generally placed at a desired location on a metal frame or thin substrate, connected by Au wire, and then the entire LSI chip is sealed with resin by transfer molding or the like, so called semiconductor.

After that, a solder ball is formed on the bottom of this semiconductor package and joined to the mother substrate in a conductive state.

At this time, the solder balls are heated and melted to about 250°C and joined. In order to improve manufacturing efficiency, the solder is automatically transferred and joined with molten solder. This process is generally called the solder reflow process. In such a process, heat of about 250°C is applied to the semiconductor package, and large thermal stress and warp deformation occur inside the semiconductor package.

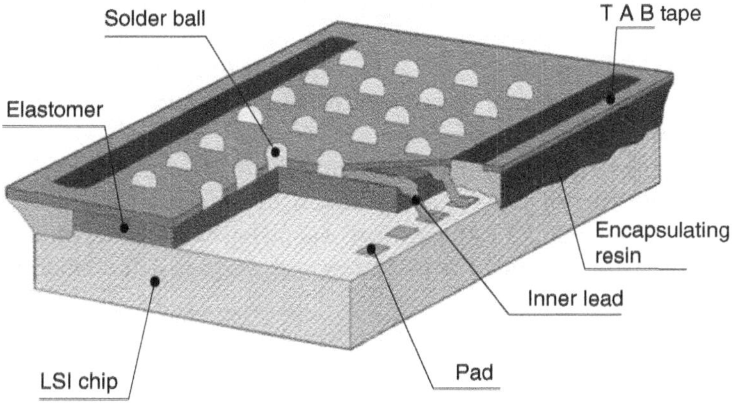

Figure 1.4 μBGA type cross-sectional structure.

Since these thermal stresses and warpage deformation cause deterioration in reliability and quality of semiconductor packages with μBGA structures shown in Fig. 1.4, it is essential to design and select the resin to be used, the manufacturing process that applies heat, and the optimized design of the μBGA structures as shown in Fig. 1.4.

That is, when various high heats are applied in the process of manufacturing an LSI package, large thermal stresses and residual stresses are generated between the constituent materials of the LSI package, which causes interfacial peeling and cracks.

At the same time, due to these thermal loads, the LSI package undergoes large thermal deformation due to differences in the linear expansion coefficient and elastic modulus of the material, causing various defects.

In recent years, wafer level package (WLP) and panel level package (PLP), which are large-scale and large-area semiconductor package and electronic device manufacturing methods, have been attracting attention for process efficiency and cost reduction.

Therefore, in addition to the optimized design of these various semiconductor packages and electronic devices, the utilization of stress/deformation simulation technology is now indispensable for shortening the development period and reducing costs. By making full use of stress/deformation simulation technology, it is possible to optimize the materials and thermal processes used for CSP packages, as well as the structure and shape, and to ensure the quality, performance, and reliability of electronic equipment. The development period can be significantly shortened, which is a great weapon for strengthening product competitiveness in each company.

When the author was engaged in research and development of semiconductor packages and electronic devices, elastic-plastic analysis by the finite element method (FEM) was the mainstream, and this analysis method was often used for resins.

The author had a failure experience. For example, there were cases where the analysis result of warp deformation by FEM was opposite to the warp direction of the real thing, and the numerical value of warp deformation was significantly different from the real thing.

With this as an opportunity, the author was keenly aware of the need for viscoelasticity analysis and carried out research. As can be seen from the photo-viscoelastic fringes shown in Fig. 1.5, the method of applying heat and the cooling rate also greatly affects the residual thermal stress and warpage deformation in the resin.

In other words, in Fig. 1.5, where the spacing between the stripes is narrow, it means that a large thermal stress remains, and from such photo-viscoelastic stripes, the location of the resin crack and the cause of warpage deformation during substrate mounting are clarified.

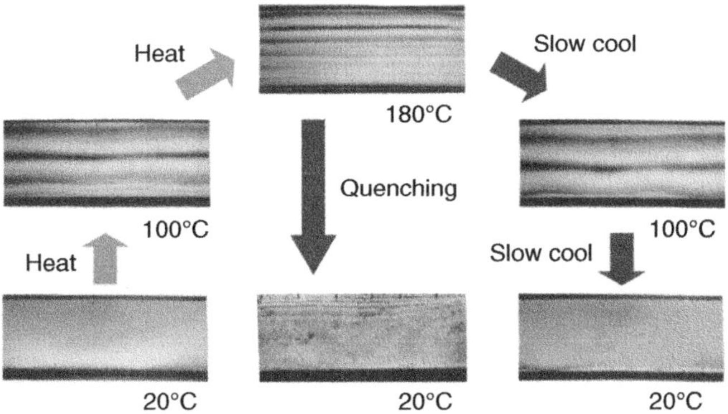

Figure 1.5 Change of thermal stress and fringe order.

From such a background and motive, the basic contents of viscoelastic mechanics are described in an easy-to-understand manner for undergraduate and graduate students of universities and those who work in companies that handle resins and at the same time, development research and technology of each company. As many practical examples as possible that are effective for product development are included in the description.

2

Basics of Polymer Materials

2.1 Polymer Material

Polymer materials (sometimes called everyday plastic materials) are organic materials that play an important role in our social life, along with inorganic materials such as metal materials and ceramics.

There are materials called polymer materials, polymers, and plastics. The definition of plastic is defined in JIS-6900, but in short, it is a molded article processed by flow using polymer as the main raw material, and rubber and elastomer are not included. A polymer substance is a substance with a large molecular weight, and a modifier, colorant, filler, reinforcing material, etc. are added to this polymer substance. Plastic is a solid that has been molded by an injection molding method using a mold die.

The raw material for plastics is naphtha, which is a product of distillation separation from petroleum. From this naphtha, a monomer that is a chemical component of the same kind and small molecule can be obtained. A polymer is composed of a large number of monomers bonded together. In other words, a large number of small molecules (monomers) gather to form a thin, strong sheet-like solid called a polymer.

A polymer is an aggregate of molecules with extremely large molecular weights and generally has a molecular weight of 10,000 or more. As a guide, ethylene, which is a monomer, is generally represented by $CH4$, and a polymer in which a large number of these are connected by a chemical reaction becomes a polymer called polyethylene (PE).

Therefore, we would like to study the outline of the types and classifications of polymer materials that are widely used in our daily life.

Polymeric Materials for Electronic Packaging, First Edition. Shozo Nakamura.

Published 2023 by John Wiley & Sons, Inc.

2.2 Types and Classification of Polymer Materials

Plastic materials are widely used in everyday miscellaneous goods, home appliances such as televisions and air conditioners, automobile parts, and aircraft parts. Furthermore, so-called fiber-reinforced plastics filled with glass fibers (CFRP) and fiber-reinforced plastics filled with carbon fibers (GFRP) are also lightweight and high-strength materials.

In recent years, with the development of the electronics industry, electronic devices and semiconductor packages used in mobile phones, laptop computers, smartphones, iPads, etc., and various electronic information devices such as mounting boards have been developed. It has also been used extensively.

Figure 2.1 shows the types and classification of resins. As shown in Fig. 2.1, resins can be broadly classified into thermoplastic resins and thermosetting resins. Among thermosetting resins, epoxy resin is generally used for large-scale integrated (LSI) packages.

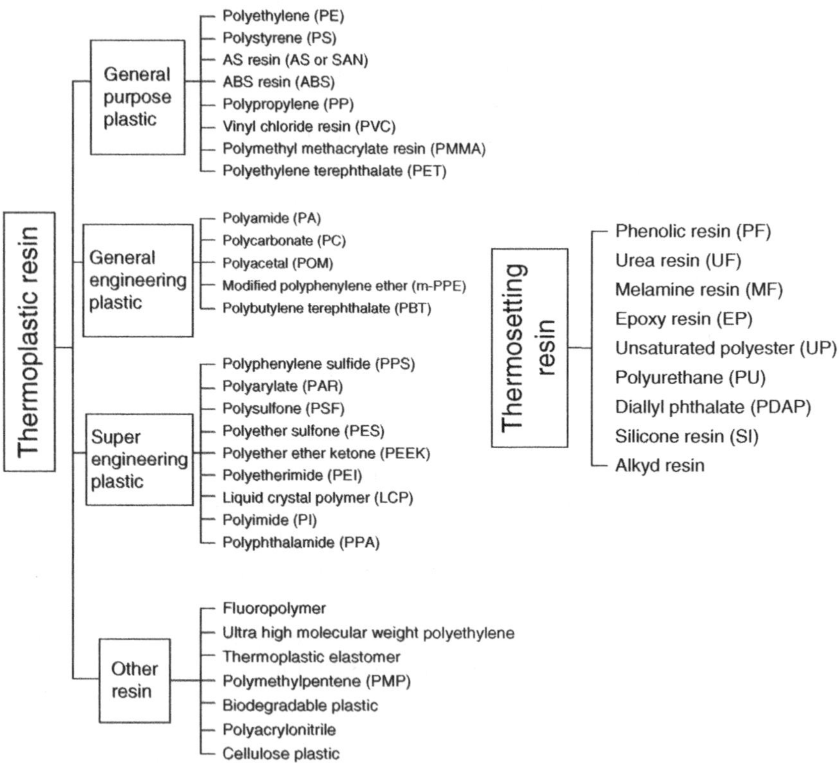

Figure 2.1 Classification and types of resin.

Thermoplastic resins are polymer substances from the beginning and melt and become fluid when heated, and substances that were of low molecules before use are heated.

There are thermosetting resins that chemically react and change into polymer substances. Most of the polymer materials used in the world are thermoplastic resins, about 85%, and the remaining 15% are thermosetting resins. Thermoplastic resins are used for parts of home appliances, while thermosetting resins are used for electronic devices such as semiconductor packages.

2.2.1 Thermoplastic Resin

Engineering plastics are general-purpose plastics such as PE and polypropylene and those with relatively excellent mechanical properties and heat resistance such as polycarbonate and polyamide. These general-purpose plastics are also thermoplastic resins and thermosetting resins.

Among them, PE is one of the cheapest and easy-to-process materials for making sheet materials such as wraps, films, and food containers by injection molding, extrusion molding, and blow molding.

Thermoplastics are simply reversible and recyclable resins that soften when heated, then melt and become fluid, and cure when cooled. The molding method is generally an injection molding method.

2.2.2 Thermosetting Resin

Thermosetting resins include epoxy resins, phenolic resins, and urea resins. Simply put, thermosetting resins are cured by the reaction when heated, and then reheated later. It is a resin that does not soften even if it is heated. Also, if it is heated to a considerably high temperature, the cross-linking bond is broken and it is carbonized and deteriorated.

Although it has excellent properties, it is brittle. The resin used for electronic devices such as semiconductor devices is this thermosetting resin, and epoxy resin is typically used as a resin for encapsulating the LSI chip inside the semiconductor package. Epoxy resin filled with silica or the like is generally used, and the transfer molding method shown in Fig. 2.2 is often used as the manufacturing method.

This molding method is presolidified on a tablet. After preheating the resin to about 90°C at high frequency, the tablet is put into the hole called the pot of the transfer molding machine, and the resin is injected through the runner into the cavity formed in the mold by the plunger. The resin held in the mold for a certain period of time cures and becomes a molded product. After opening the mold and taking out the molded product.

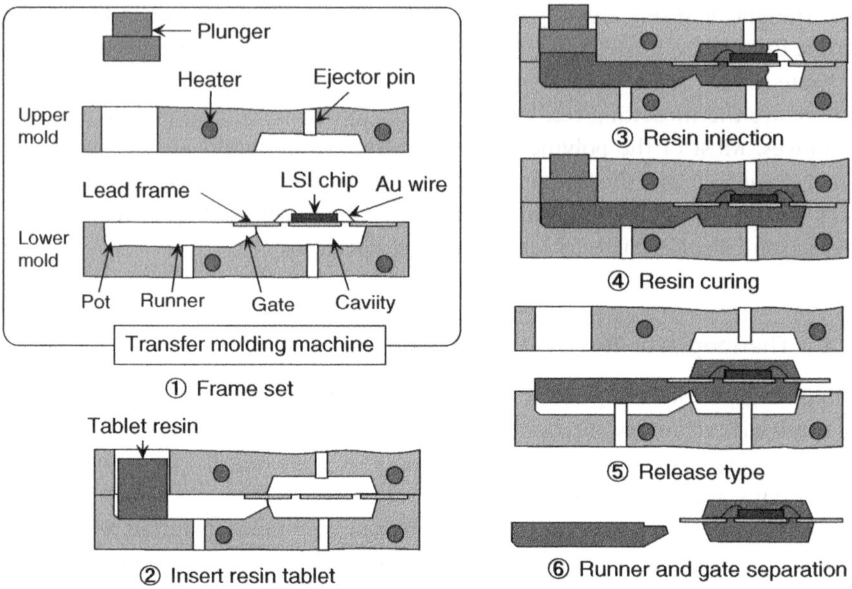

Figure 2.2 Outline of transfer molding method.

In particular, for electronic parts such as semiconductor packages and electronic devices, silica-filled epoxy resins in various forms are often used depending on the application.

2.3 General Properties of Polymer Materials

Resin materials are significantly inferior in mechanical strength and thermal properties compared to metals and ceramics. On the other hand, they are lightweight and inexpensive and suitable for mass production. Also, taking advantage of their high price but light weight. Composite materials such as fiber-reinforced plastics filled with glass fibers (CFRP) and carbon fibers (GFRP) are also being used more and more as lightweight and high-strength materials.

The features of plastic materials that will be described in detail in chapters 5 and 6 are stress relaxation and creep phenomena. That is, the thermal and mechanical properties of plastic materials are greatly affected by time and temperature, that is, the properties of viscoelasticity.

In the range where the deformation is relatively small, the mechanical behavior at low temperature is elastic, so-called Hooke's law, $\sigma = E\varepsilon$ holds, the stress σ is proportional to the strain ε, and the proportionality constant E is the elastic modulus.

This relationship is usually expressed by "spring." On the other hand, at high temperature, viscoelastic behavior is prioritized, Newton's law, $\boldsymbol{\sigma} = \boldsymbol{\eta}\frac{\mathbf{d}\boldsymbol{\varepsilon}}{\mathbf{d}t}$. The proportionality constant η, which is proportional to the velocity $\mathbf{d}\boldsymbol{\varepsilon}/\mathbf{d}\boldsymbol{t}$, is the viscosity coefficient or viscosity. This relationship is usually expressed by a "dashpot" or damper.

From another point of view, if the temperature dependence of the elastic modulus E of the plastic material is investigated, the region where the elastic modulus is generally large at low temperature is the region viscoelastic or leather-like where the elastic modulus decreases as the temperature rises. The region changes to a rubber-like region that becomes soft through the region.

In the case of a thermoplastic resin, it changes from a rubber-like state to a viscoelastic flow state and changes to a liquid. The temperature that changes (softens) is generally called the glass transition temperature.

In semiconductor packages used for electronic devices as described in Chapter 7, they are sealed with resin to protect the LSI chip, but the viscoelastic properties of the resin used here are the warp deformation and strength of electronic devices such as semiconductor packages. This is a particularly important issue because it greatly affects reliability.

2.4 Summary

Polymer materials (plastic materials) are used in a variety of applications from everyday products to aircraft. There are innumerable types and classifications. Typical examples are thermoplastic resins and thermosetting resins.

These resins are used properly according to their respective uses and purposes. The properties of any polymer material are significantly different from those of metals and ceramics.

The characteristic is "viscoelasticity," which is a material whose properties differ greatly depending on time and temperature. Among these polymer materials, epoxy resins containing a large amount of fillers such as silica are often used for electronic devices such as semiconductor packages in the electronics field.

These epoxy resins are also used in recent electronic devices and mounts. Materials such as epoxy resins tend to change drastically due to the development of substrate manufacturing methods.

In any case, since resins, which are organic materials, have complicated viscoelastic behavior, they are highly reliable and have high performance. It goes without saying that the optimized design of the material, structure, and thermal process is extremely important for making electronic products such as devices.

Therefore, when developing and designing various electronic products, not only shortening the development period, but also reducing the manufacturing cost.

In order to reduce manufacturing costs, stress/deformation simulation technology that fully understands the viscoelastic properties of resins is indispensable. In this book, after first understanding the concept of what a polymer material is, and then understanding the types and basics of classification, the necessity and importance of viscoelasticity analysis in product development will be described based on specific cases. Please refer to other books regarding the synthesis of polymer materials and chemical reactions.

3

Basics of Elastic Theory

3.1 Elasticity

This book describes viscoelastic analysis of strength and warpage deformation for the purpose of highly reliable design of large-scale integrated (LSI) packages and optimized design of resin materials.

There are many analysis software about the strength and deformation in the world, but most of them are based on elastic theory.

In recent years, analysis considering viscoelasticity has been gradually carried out, but from the author's experience, it cannot be said to be sufficient.

However, before talking about viscoelasticity, it is necessary to know the basics of elasticity theory. From that point of view, I will touch on the basics of elasticity theory.

In general, the finite element method (FEM) is the most frequently used numerical analysis method as a simulation technique in the strength design and structural design of machines and structures. This analysis by the FEM (hereinafter abbreviated as FEM analysis) is based on the elasticity theory as will be explained later in this chapter.

In order to understand viscoelasticity, its analysis method, and the basic formulas and formulation of viscoelasticity described in this book, it is very useful to first know the basic formulas and concepts of elastic theory. From this point of view, the basics of elasticity theory and FEM analysis will be briefly described next.

3.2 Stress and Strain

Stress is the value obtained by dividing the load applied to an object by the unit area. Strain is the value obtained by dividing the stretched (or shrunk) dimension by the original dimension. That is, the force per unit area generated in an object

Polymeric Materials for Electronic Packaging, First Edition. Shozo Nakamura.

Published 2023 by John Wiley & Sons, Inc.

with respect to the load is the stress. There are two types of strain, one is longitudinal strain (ε) and the other is lateral strain (ε'), and the ratio (ε'/ε) is called Poisson's ratio ν.

Now, in FEM analysis based on elastic theory, the following assumptions are generally made.

1) The object to be analyzed is a completely elastic body. In other words, it deforms when it is loaded on an object, but it has the property of completely restoring to its original shape when the load is removed.
2) The material is homogeneous and isotropic. That is, it has the same properties no matter where it is taken, and there is no anisotropy.
3) The strain when loaded on an object is small (deformation is also minute), and there is a proportional relationship.

Generally, when an external force is applied to an object, stress σ is generated inside the object. Now, considering a cube element, three types of normal stress components and six types of shear stress components parallel to the surface are generated on each surface of the cube.

That is, three normal stresses, σ_x, σ_y, σ_z; six shear stresses, τ_{xy}, τ_{yx}, τ_{yz}, τ_{zy}, τ_{zx}, τ_{xz}; and these stresses are a total of nine stress vectors with magnitude and direction. Here, since the cube elements are in a balanced state, $\tau_{xy} = \tau_{yx}$, $\tau_{yz} = \tau_{zy}$, $\tau_{zx} = \tau_{xz}$ holds, and the total stress is six in the end.

3.3 Finite Element Method Analysis (FEM Analysis)

3.3.1 What Is FEM Analysis?

As mentioned above, when an external force is applied to an object, stress, strain, and deformation occur inside the object. FEM analysis is a numerical analysis method for determining the stress, strain, and deformation generated inside a structure when various external forces are applied to the structure. This method has been recognized as a numerical analysis method for industrial problems with the rapid improvement in the performance of computers and has greatly contributed to the rise of computational mechanics in recent years.

It was first used in the 1950s for stress analysis in structural mechanics by the matrix method. FEM analysis has been used exclusively in the field of strength calculation of large structures such as aircraft and ships, but it has been found that this is an analysis method for partial differential equations, and the fields of application are also electromagnetic fields, heat conduction, and fluids. It has been expanded to analyze a wide range of physical phenomena and engineering fields as a simulation of air flow and the like.

3.3.2 Concept of FEM Analysis

The concept of FEM analysis is explained. First, the area of the object to be analyzed is divided into elements. There are many types of elements.

Next, each element is formulated by the energy principle, etc., and the element rigidity equation for each element is derived. Then, the stress, strain, and displacement of the angle (node) of each element are obtained by collecting the stiffness equations of each element over all regions and finally solving the simultaneous equations.

That is, the amount of deformation of the object can be known from the amount of movement of the node coordinates before and after the external force. When dividing the area of an object into elements, the shape of the object is inevitably approximated.

In order to increase the accuracy of the area, it is possible to easily increase the number of element divisions with automatic mesh division software, but on the other hand, if the number of division elements and the number of nodes increase infinitely, preparation before analysis and since the author has also experienced that there is a limit to the calculation accuracy for the complicated calculation, it is desirable to use an appropriate number of divisions.

In the actual FEM analysis, there are roughly three steps. First, the following conditions are determined in the preparatory stage of preprocessing.

i) Input element division data such as the shape and dimensions of the object;
ii) material property values such as elastic modulus, linear expansion coefficient, and Poisson's ratio;
iii) restraint conditions; and
iv) load conditions such as heat and external force.

Next, FEM analysis is performed based on the above elasticity theory and equations. Many analysis software such as ADINA, MARC, and ANSYS are commercially available. Finally, as post-processing, the three analysis results will be obtained.

① Node displacement
② Strain of node
③ Stress of node

In any case, FEM analysis has advantages and disadvantages, and it is not suitable for materials showing viscoelastic behavior such as resin, so it is necessary to firmly recognize the limits of FEM analysis and the purpose of use.

3.4 Governing Equation of Elastic Body

The relationship between the stress component, strain component, and displacement component can be divided into the following three equations.

$$\frac{\partial \sigma_{xx}}{\partial x} + \frac{\partial \tau_{yx}}{\partial y} + \frac{\partial \tau_{zx}}{\partial z} + F_x = 0 \tag{3.1}$$

$$\frac{\partial \tau_{xy}}{\partial x} + \frac{\partial \sigma_{yy}}{\partial y} + \frac{\partial \tau_{zy}}{\partial z} + F_y = 0 \tag{3.2}$$

$$\frac{\partial \tau_{xz}}{\partial x} + \frac{\partial \tau_{yz}}{\partial y} + \frac{\partial \sigma_{zz}}{\partial z} + F_z = 0 \tag{3.3}$$

3.4.1 Balance Equation of Force and Moment

The following equation holds for nine stress components: σ_x, σ_y, σ_z, τ_{xy}, τ_{yx}, τ_{yz}, τ_{zy}, τ_{zx}, τ_{xz}.

Here, F_x, F_y, and F_z are body forces acting on a minute cube and are components in the x, y, and z directions.

Although omitted here, these equilibrium equations are equivalent to the principle of virtual work. In a nutshell, the principle of virtual work is that an object deforms when an arbitrary load is applied to an elastic body of any shape. It shows that the external virtual work due to the virtual displacement given to this deformed object is equal to the increment of internal energy, that is, the internal virtual work.

3.4.2 Strain–Displacement Relational Expression

Deformation and strain occur when a load acts on an object. Deformation has a displacement component, and a strain component also occurs. If the displacement components in the x, y, and z directions are u, v, and w, respectively, and the strain components are ε_x, ε_y, ε_z, γ_{xy}, γ_{yz}, and γ_{zx}, they have the following relationship.

$$\varepsilon_x = \frac{\partial u}{\partial x} \tag{3.4}$$

$$\varepsilon_y = \frac{\partial v}{\partial y} \tag{3.5}$$

$$\varepsilon_z = \frac{\partial w}{\partial z} \tag{3.6}$$

$$\gamma_{xy} = \frac{\partial v}{\partial x} + \frac{\partial u}{\partial y} \tag{3.7}$$

$$\gamma_{yz} = \frac{\partial w}{\partial y} + \frac{\partial v}{\partial z} \tag{3.8}$$

$$\gamma_{zx} = \frac{\partial u}{\partial z} + \frac{\partial w}{\partial x} \tag{3.9}$$

3.4.3 Stress–Strain Relational Expression

There is the following relationship between stress and strain.

$$\varepsilon_x = \frac{1}{E}\{\sigma_x - \nu(\sigma_y + \sigma_z)\} \tag{3.10}$$

$$\varepsilon_y = \frac{1}{E}\{\sigma_y - \nu(\sigma_z + \sigma_x)\} \tag{3.11}$$

$$\varepsilon_z = \frac{1}{E}\{\sigma_z - \nu(\sigma_x + \sigma_y)\} \tag{3.12}$$

$$\gamma_{xy} = \frac{2(1+\nu)}{E}\tau_{xy} \tag{3.13}$$

$$\gamma_{yz} = \frac{2(1+\nu)}{E}\tau_{yz} \tag{3.14}$$

$$\gamma_{zx} = \frac{2(1+\nu)}{E}\tau_{zx} \tag{3.15}$$

Here,

$$G = \frac{E}{2(1+\nu)} \tag{3.16}$$

G: lateral elastic modulus, or shear elastic modulus.

From the above, a total of 15 unknowns, which are 6 stress components, 6 strain components, and 3 displacement components, can be obtained by 15 equations.

3.5 Law of Elastic Breakage

There are the following theories as to the conditions under which the elastic body is damaged.

3.5.1 Yield Conditions

When the stress values reach any combination, the elastic behavior, that is, the linearity of stress and strain, is lost in the ductile material, and the mechanical conditions indicating whether the plastic deformation, that is, yields, or the fracture in the brittle material, are determined. It is called an elastic failure criterion or a yield criterion.

3.5.2 Maximum Principal Stress Theory

The theory is that when any of the three principal stresses of the x-axis, y-axis, and z-axis generated inside the material reaches the limit value, it will be damaged.

Now, when the principal stress is $\sigma_1 > \sigma_2 > \sigma_3$,

$$\sigma_1 = \sigma_y \tag{3.17}$$

is the condition of damage. Here, σ_y is the yield stress at the time of simple tension or simple compression. This theory is suitable for ductile materials but not for brittle materials.

3.5.3 Maximum Main Strain Theory

The theory is that the maximum principal strain generated inside the material will be damaged when $\varepsilon_y = \sigma_y/E$ with the strain ε_y corresponding to the yield stress σ_y.

Now, when the strain is $\varepsilon_1 > \varepsilon_2 > \varepsilon_3$,

$$\varepsilon_1 = \frac{\sigma_1 - \nu(\sigma_2 + \sigma_3)}{E} = \varepsilon_y = \frac{\sigma_y}{E} \tag{3.18}$$

is the condition of damage.

3.5.4 Maximum Shear Stress Theory

This theory is also called Tresca's conditional expression. If the principal stress in the triaxial state is $\sigma_1 > \sigma_2 > \sigma_3$, and the maximum shear stress

$$\tau_{\max} = \frac{\sigma_1 + \sigma_2}{2} \tag{3.19}$$

is the yield stress σ_y of simple tension, then

$$\tau_{\max} = \sigma_y/2 \tag{3.20}$$

That is,

$$\tau_{\max} = \frac{\sigma_1 - \sigma_2}{2} \tag{3.21}$$

is the condition of damage.

3.5.5 Shear Strain Energy Theory

This theory is also called Mises' conditional expression.

The surrender condition is

$$\overline{\sigma} = \frac{\sqrt{(\sigma_1 - \sigma_2)^2 + (\sigma_2 - \sigma_3)^2 + (\sigma_3 - \sigma_1)^2}}{2} = \sigma_y \tag{3.22}$$

Here, $\overline{\sigma}$: equivalent stress, σ_y: yield stress of simple tension.

In the case of simple tension, because $\sigma_2 = \sigma_3 = 0$,

$$\overline{\sigma} = \sigma_1 = \sigma_2 \tag{3.23}$$

3.5.6 Elastic Strain Energy Theory

The theory is that the strain energy per unit volume at a certain point in the material will be damaged when it reaches a certain value peculiar to the material.

Here, the principal stress is σ_1, σ_2, σ_3, and if the maximum elastic energy of simple tension is $\boldsymbol{\sigma_y}^2/2\boldsymbol{E}$, the yield condition U is shown as the following equation:

$$U = \frac{\sigma_1{}^2 + \sigma_2{}^2 + \sigma_3{}^2 - 2\nu(\sigma_1\sigma_2 + \sigma_2\sigma_3 + \sigma_3\sigma_1)}{2E} = \frac{\sigma_y{}^2}{2E} \tag{3.24}$$

3.6 Plane Stress and Plane Strain

In Section 3.4, we mainly discussed the three-dimensional stress field. However, in stress analysis, it is not necessary to analyze everything as a three-dimensional problem, and in some cases it is sufficient to analyze it as a two-dimensional stress problem.

For reference, Fig. 3.1 is a schematic representation of the three-dimensional stress field and the two-dimensional stress field. Here, in two-dimensional stress, it is necessary to distinguish between plane stress and plane strain in the numerical analysis.

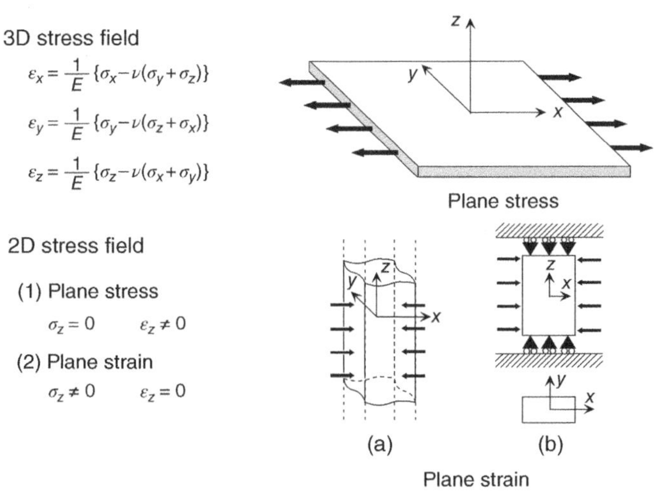

Figure 3.1 Stress analysis field for numerical analysis.

4

Stress Evaluations with Defects

4.1 Difference from Strength of Materials

The stress and strain handled in strength of materials are treated as if the entire material is homogeneous and continuous with no anisotropy or cracks. However, in general, any object has defects such as abrupt shape changes such as notches and minute cracks. When discussing safe strength design such as stress and strain of a structure, it is necessary to consider these notches and micro-cracks.

This is because the defects such as notches and cracks in these materials are extremely dangerous, large stresses are concentrated near these defects, and the object may be destroyed at once with a small external force. The discipline dealing with such cracks and defects is fracture mechanics, which cannot be dealt with by the concept of strength of materials.

Considering that the electronic components such as semiconductor packages mentioned in this document are also composed of a structure in which a large-scale integrated (LSI) chip is sealed with a resin and different materials such as a lead frame, a solder ball, and a printed circuit board, a bonding interface exists. In particular, if a thermal external force acts on the bonding interface between a resin with low strength and an LSI chip or a resin and a metal lead frame, a large thermal force will be applied to the bonding interface due to the difference in elastic coefficient and linear expansion coefficient between the two materials. Shear stress was generated, and there was a risk of problems such as interfacial peeling and interfacial cracking.

In fact, in the semiconductor package the author was involved in around 1980, there were problems such as cracks in the resin due to heat load and peeling at the adhesive interface, but nowadays when the structure and mounting technology of the semiconductor package have advanced, it is destroyed. The need for mechanical strength analysis has almost disappeared.

Polymeric Materials for Electronic Packaging, First Edition. Shozo Nakamura.

Published 2023 by John Wiley & Sons, Inc.

In addition, the problems of recent semiconductor packages and mounting methods are not so much the problems caused by the structure of cracks and defects, but rather the problems related to the viscoelastic properties of the resin.

Therefore, in this book, the basic contents of stress concentration and fracture mechanics are described.

4.2 Stress Concentration and Stress Intensity Factor

4.2.1 Stress Concentration

The stress distribution is not uniform in places where the cross-sectional area changes suddenly, such as when there are notches in the object, and large stresses occur in the parts where the shape changes suddenly. This phenomenon is called stress concentration, and fracture or deformation occurs from this part. In the actual design, it is necessary to design so that stress is not concentrated. First, in order to understand the stress concentration in an easy-to-understand manner, let us look at the difference in stress applied to the test piece in Fig. 4.1.

Obtain the stress when the member in Fig. 4.1 is pulled in the vertical direction with the load P. The stress is the load P divided by the cross-section A, and $\sigma =$ load P/cross-section A.

In Fig. 4.1a, uniform stress is generated in the cross-section A, but if there is a notch in a part of the shape as in the case of Fig. 4.1b, the stress is concentrated in the vicinity and the average stress in (a).

In (b), a much larger $\sigma_{\max}$ is generated with respect to σ_0. Here, if the degree of stress concentration is expressed by the stress concentration coefficient α,

$$\alpha = \frac{\sigma_{\max}}{\sigma_0} \tag{4.1}$$

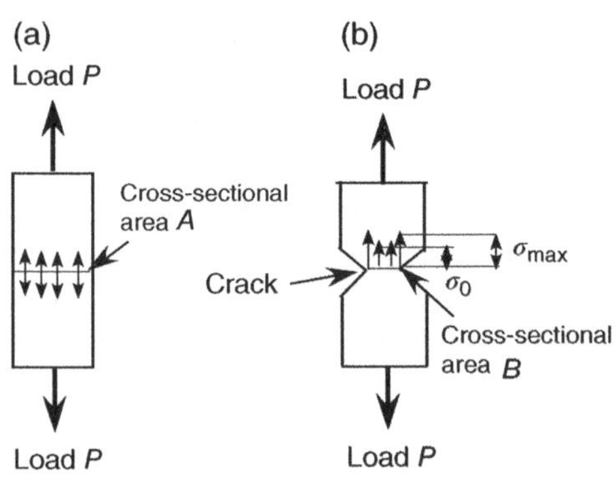

Figure 4.1 Stress states with and without cracks. (a) Test piece with a uniform cross section. (b) Schematic diagram showing a case with cracks.

4.2.2 Stress Intensity Factor

In general, the stress intensity factor is a physical quantity that represents the strength of the stress distribution near the crack tip, which is derived by linear elastic mechanics. It is one of the basic physical quantities of fracture mechanics and is used to evaluate the strength of materials with cracks and defects.

If there is a defect such as a crack in the member, the stress in that part becomes extremely large, the tip of the crack becomes a stress singular field, and the stress cannot be specified.

Therefore, in order to represent such a stress singular field near the crack, it is necessary to evaluate using the stress intensity factor. As shown in Fig. 4.2, consider a case where a uniform tensile stress σ acts on an infinitely large flat plate having a crack having a length of $2a$ at a sufficiently distant place. σ_y becomes infinite as it approaches the crack tip ($x \to a$).

On the other hand, on the crack surface ($x < a$), $\sigma_y = 0$, and the crack tip is a stress singular point, which is called the stress specificity or singular stress field of the crack tip.

The stress σ_y in the y direction is expressed by the following equation (4.2).

$$\sigma_y = \sigma \frac{x}{\sqrt{x^2 - a^2}} \tag{4.2}$$

Considering a local coordinate system in which the origin O' is moved to the tip of the crack. Figure 4.3 shows the horizontal axis of the local coordinate system as the r-axis.

Since $r = 0$ at $x = a$, substituting $x = (r + a)$ into Eq. (4.1), σ_y is expressed by Eq. (4.2).

In addition, the following equation is obtained by expanding Eq. (4.3) into series. It becomes the following equation.

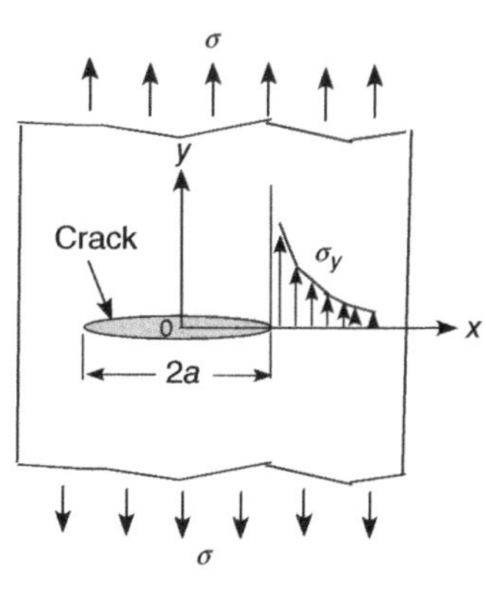

Figure 4.2 Stress of an infinitely large flat plate with cracks (coordinate origin is the center of the crack).

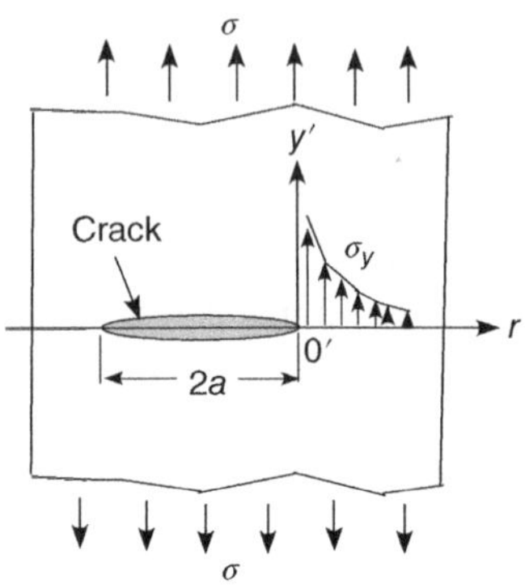

Figure 4.3 Stress of an infinitely large flat plate with cracks (coordinate origin is the edge of the crack).

In $r \ll a$, the first term of Eq. (4.3) is overwhelmingly larger than the other terms, so if the second and subsequent terms are ignored, σ_y becomes the following equation (4.4).

$$\sigma_y = a\left\{\sqrt{\frac{a}{2r}} + \frac{3}{4}\sqrt{\frac{r}{2a}} - \frac{5}{16}\left(\sqrt{\frac{r}{2a}}\right)^3 + \cdots\right\} \quad (r \geq 0) \tag{4.3}$$

$$\sigma_y = \sigma\sqrt{\frac{a}{2r}} \quad (r \geq a) \tag{4.4}$$

The stress near the crack is a stress singular field that changes in proportion to $1/\sqrt{r}$. Substituting numerical values into Eqs. (4.2) and (4.4) and calculating the difference, it is about 7% when r/a is 1/10, so the approximation of Eq. (4.4) holds. R is about 1/20 of the crack length. We will transform Eq. (4.4).

Here, K is given by the following formula.

$$\sigma_y = \sigma\sqrt{\frac{a}{2r}} = \sigma\sqrt{\frac{\pi a}{2\pi r}} = \frac{K}{\sqrt{2\pi r}} \quad (r \geq a) \tag{4.5}$$

And, K is stress intensity factor.

$$K = \sigma\sqrt{\pi a} \tag{4.6}$$

K is like the proportionality constant of the stress singularity of $1/\sqrt{r}$ and is called the stress intensity factor, which is the most important parameter in fracture mechanics. The unit of stress intensity factor is $(\mathrm{MPa}\sqrt{\mathrm{m}})$.

5

Basics of Viscoelasticity

5.1 About Viscoelasticity

There are a wide variety of materials for making electronic equipment, from large structures seen in daily life to precision electronic devices. In particular, precision equipment such as electronic equipment includes semiconductor packages and electronic components that require precision on the order of nanometers to microns. These cause defects such as large warp deformation and interface peeling even with a small heat load or external force.

Therefore, in product design of large-scale integrated (LSI) packages, it is indispensable to optimize the constituent materials, structure and shape, and the thermal process during manufacturing from the viewpoint of warpage deformation and strength. Many LSI packages and electronic components are used in electronic devices, and the thermal and mechanical properties of resins greatly affect the reliability of products.

Unlike metal and ceramic materials, resins have complicated and troublesome properties that are greatly affected by temperature and time, and these properties are called viscoelasticity.

Many resins are used in semiconductor packages such as chip scale packages (CSPs) and electronic components such as motherboards and printed circuit boards. Generally, an epoxy resin filled with a large amount of filler such as silica (SiO2) is used as a sealing material for protecting the LSI chip.

In addition, a semiconductor package is melt-bonded to a printed circuit board (in general FR-4) with a large number of solder balls, and resin materials such as elastomer and underfill are also used for stress buffering at the solder balls. Thermosetting resin is also used for FR-4 substrates in which epoxy resin is dipped in glass fiber or glass cloth and laminated in multiple layers. In this way, a wide variety of resins are used for electronic components, and since these complicated and

Polymeric Materials for Electronic Packaging, First Edition. Shozo Nakamura.

Published 2023 by John Wiley & Sons, Inc.

troublesome properties greatly affect the reliability of products, it is necessary to design with an accurate understanding of the viscoelastic properties of the resin.

This chapter basically studies the properties of resin materials with viscoelastic properties and describes the basics of viscoelasticity such as simple mechanical models and formulations that express viscoelasticity.

5.2 Elasticity, Viscosity, and Viscoelasticity

As shown in Fig. 5.1, elasticity is the property that an object undergoes appropriate deformation when an external force is applied to it and returns to its original state when it is unloaded. An object with elastic properties is called an elastic body.

That is, if the stress σ and the strain ε are in a proportional relationship and the constant of proportionality is the elastic modulus E, then Hooke's law

$$\sigma = E\varepsilon \tag{5.1}$$

And this elasticity is generally expressed as a "spring." Viscosity is the property due to which, as shown in Fig. 5.1, a corresponding deformation occurs when an external force is applied, but the deformation does not recover to the original state even when the load is removed, and permanent deformation occurs. An object with this property is viscous.

That is, the stress σ and the strain rate $\mathrm{d}\varepsilon/\mathrm{d}t$ are in a proportional relationship, and if the proportionality constant is the viscosity coefficient (or viscosity) η, it is ideally Newton's law.

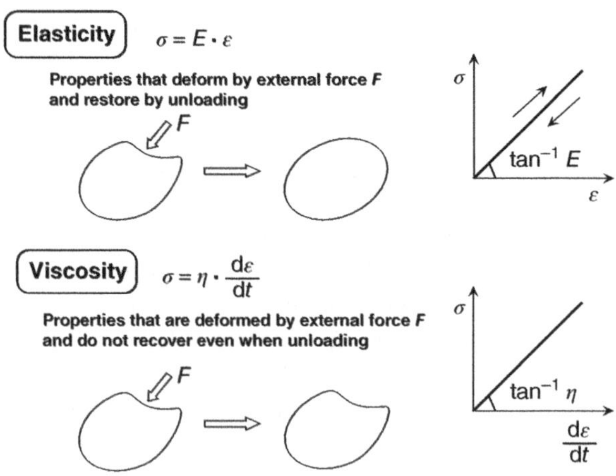

Figure 5.1 Schematic diagram of elasticity and viscosity.

$$\sigma = \eta \frac{d\varepsilon}{dt} \tag{5.2}$$

And the viscous behavior is generally represented by a syringe-like "dashpot" (or damper). If you imagine the magnitude of this viscosity coefficient, for example, water has a small viscosity coefficient, so it can easily flow by applying a small force, but on the other hand, starch syrup and mayonnaise have a large viscosity coefficient, which means that it is difficult to flow.

On the other hand, viscoelasticity is a property that has both elasticity and viscosity, and its details are expressed by a mechanical model in which a spring representing elasticity and a dashpot representing viscosity coexist, as will be described later. Maxwell model and Voigt model are famous as typical mechanical models.

In particular, viscoelastic materials such as resins have more complicated and esoteric problems in material design because the thermal and mechanical properties of the resins are greatly affected by time and temperature.

5.3 Stress and Strain Response

When a constant stress σ (external force) is applied to an object, strain ε (deformation λ) occurs. Figure 5.2 shows the mechanical behavior of solid body.

That is, it is a relational expression between elastic and viscous stress and strain and strain rate. In addition, their mechanical behavior is schematically shown. Polymer materials and elastomers are characterized by showing a hysteresis curve.

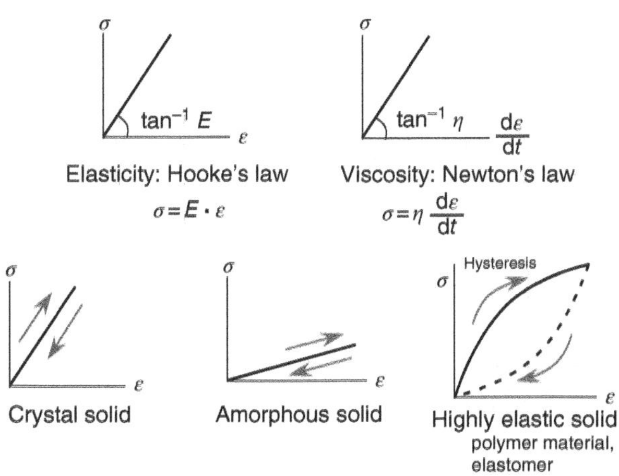

Figure 5.2 Mechanical behavior of solid body.

Figure 5.3a,b schematically shows the mechanical behavior of a solid, similar to Fig. 5.2. That is, elasticity behavior includes energy elasticity (or glassy elasticity) and entropy elasticity (or rubbery elasticity). In both cases, the temporal change in strain under constant stress is constant.

On the other hand, in the case of viscoelasticity, strain increases with the passage of time under constant stress. This phenomenon is called the creep phenomenon, which is a major feature of viscoelasticity as described later.

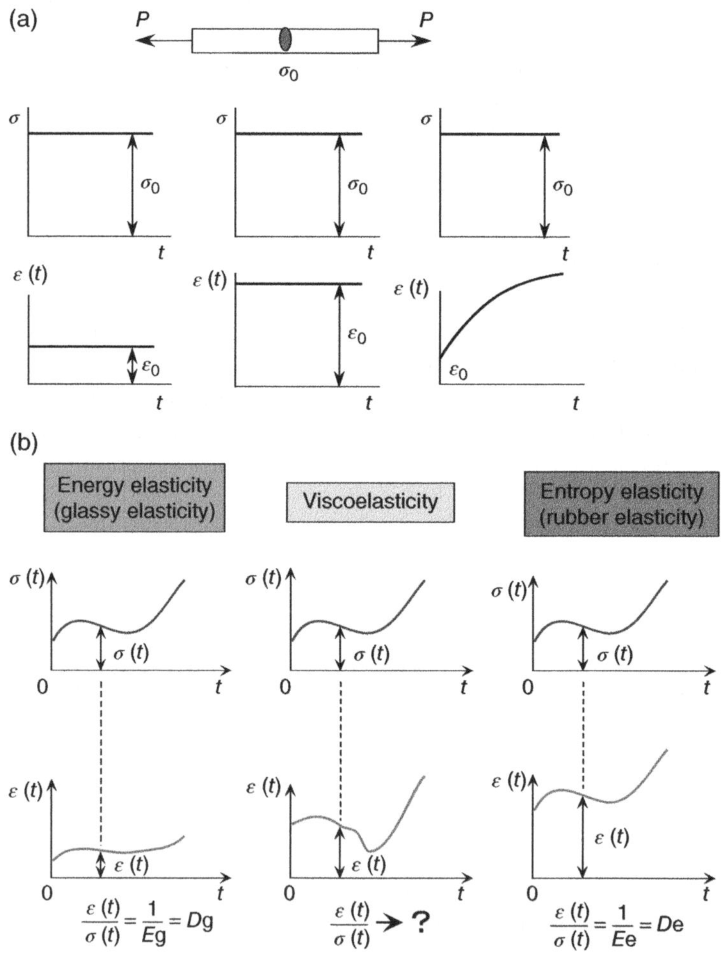

Figure 5.3 (a) Elastic and viscoelastic mechanical behavior. (b) Mechanical characteristics of viscoelasticity.

Now, the creep phenomenon is explained in Fig. 5.4. When a constant load W is applied to a test piece having a length L, the test piece expands with the passage of time. Here, the mechanical model before loading is expressed by the Voigt model. Immediately after loading, the test piece is instantly elastically deformed, and in the model, spring A is stretched (ε_1). After that, the dashpots A and B gradually grow with the passage of time and show a constant growth (strain) after t hours.

Thus, the Vogt model is suitable for explaining the creep phenomenon.

On the other hand, the stress relaxation phenomenon is explained in the upper Fig. 5.5. When a constant load W is applied to the test piece, the spring expands instantly, and if this elongation amount is constrained to be constant, the applied load decreases with the passage of time. Explaining this with the Maxwell model at the bottom of Fig. 5.5, the dashpot does not react immediately after loading and only the spring expands, and after that, the dashpot gradually expands and the spring contracts.

That is, the load and stress decrease with the passage of time, which is a stress relaxation phenomenon, and the Maxwell model is suitable for explaining this.

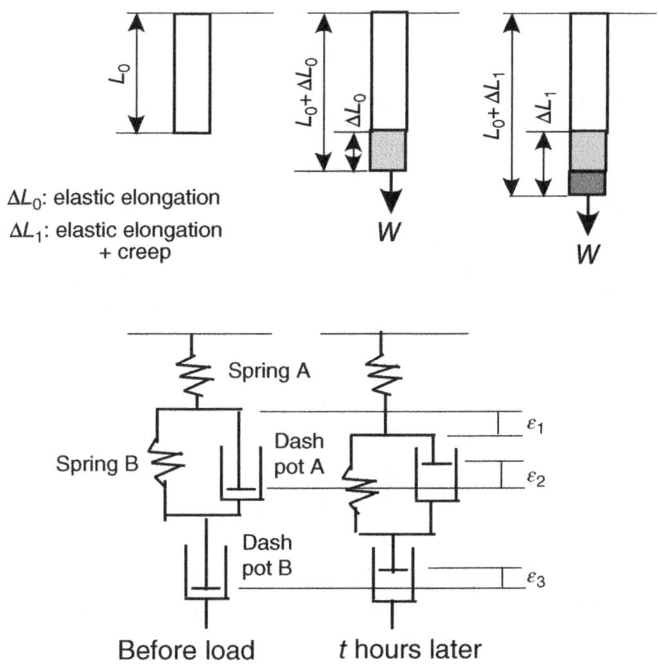

Figure 5.4 Creep phenomenon of viscoelastic material.

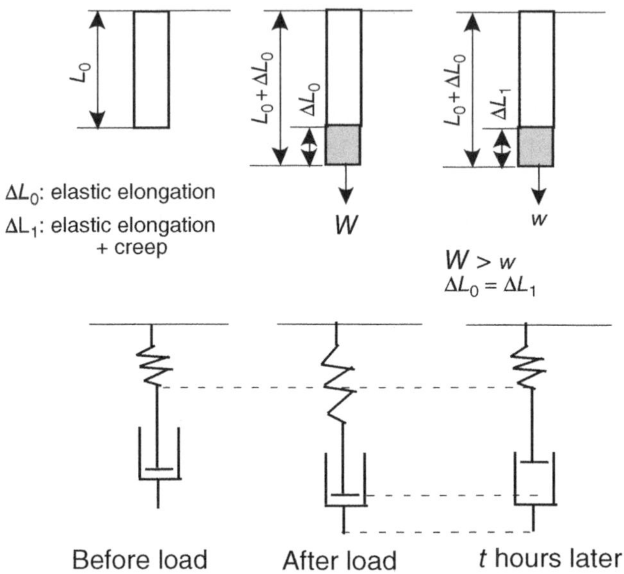

Figure 5.5 Stress relaxation phenomenon and Maxwell model.

5.4 Mechanical Model Representing Viscoelastic Properties

Viscoelasticity is a property in which both elasticity representing a solid and viscous property representing a liquid coexist as described earlier. As for elasticity, stress σ and strain ε are in a proportional relationship, and if the proportionality constant is the elasticity coefficient E, the relationship between the two is the Hooke's law that can be expressed as shown in Eq. (5.1).

On the other hand, for viscosity, stress σ and strain rate $d\varepsilon/dt$ are in a proportional relationship, and if the proportionality constant is the viscosity coefficient η, the relationship between the two is expressed by Newton's law shown in Eq. (5.2).

Figure 5.6 shows a viscoelastic dynamics model in which the elasticity shown by the spring and the viscous property shown by the dashpot coexist. Figure 5.6a shows a Maxwell model in which a spring and a dashpot are configured in series, and Fig. 5.6b shows a Voigt model in which both are configured in parallel.

First, the concept of linear viscoelastic behavior will be explained using the Maxwell model in Fig. 5.6a.

The spring indicates a solid with an elastic modulus E, and the dashpot indicates a liquid with a viscosity coefficient η, which are configured in series. When a

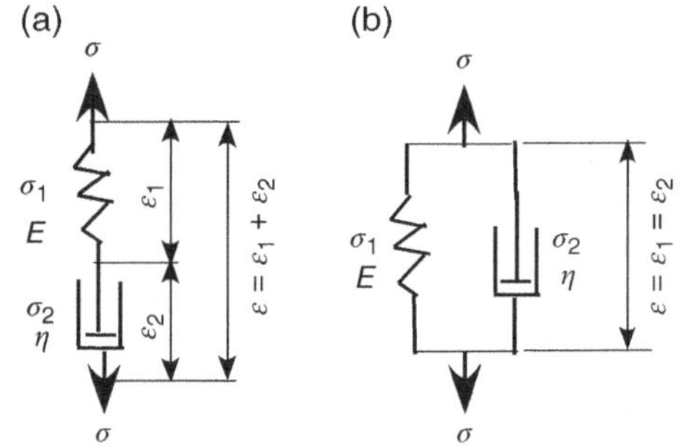

Figure 5.6 (a) Maxwell model and (b) Voigt model.

constant stress σ is applied to both ends, the stress σ_1 acting on the spring and the stress σ_2 acting on the dashpot are equal.

On the other hand, the total strain ε can be expressed as the sum of the strain ε_1 generated in the spring and the strain ε_2 generated in the dashpot. The following is a representation of these series of relationships as an expression.

$$\sigma = \sigma_1 + \sigma_2 \tag{5.3}$$

$$\varepsilon = \varepsilon_1 + \varepsilon_2$$

Here, $\sigma_1 = E\varepsilon_1$

$$\sigma_2 = \eta \frac{d\varepsilon_2}{dt} \tag{5.4}$$

σ is derived from these equations,

$$\sigma = \sigma_0 \exp\left(-\frac{E}{\eta}t\right) \tag{5.5}$$

Here, σ_0: initial stress, E: elastic modulus, η: viscosity coefficient, and t: time.

From this, it can be seen from the equation that the Maxwell model is a mechanical model suitable for explaining the stress relaxation phenomenon as described above.

Similarly, in the Voigt model in which the spring and dashpot in Fig. 5.6b are configured in parallel, the total strain ε is equal to the spring strain ε_1 and the dashpot strain ε_2 ($\varepsilon = \varepsilon_1 = \varepsilon_2$).

On the other hand, the stress σ can be expressed as the sum of the stress σ_1 and σ_2. Now we derive the equations for these in the same way as the Maxwell model,

$$\varepsilon = \frac{\sigma}{E}\left\{1 - \exp\left(-\frac{E}{\eta}t\right)\right\} \tag{5.6}$$

It can be seen that the Voigt model is a mechanical model suitable for explaining the creep phenomenon, which is another feature of viscoelastic behavior.

For reference, the combined phenomenon of stress relaxation and creep phenomenon of viscoelastic body are explained by the mixed mechanical model composed of both Maxwell model and Voigt model in Fig. 5.7a,b.

Figure 5.7a shows general creep characteristics. When a load is applied to an object, a strain response of ε_1 occurs instantly, and then the strain increases with the passage of time. It is a schematic diagram in which the strain instantly decreases by ε_1 due to unloading, then gradually decreases with the passage of time, and stabilizes at ε_2.

Figure 5.7b expresses this series of behaviors of creep characteristics with a mixed mechanical model consisting of a Maxwell model and a Voigt model.

That is, the elastic spring E_1 responds instantaneously to the elastic strain of ε_1 due to the load of the external force, and then the viscous dashpot η_1 responds with a time delay to generate the strain ε_2 due to the viscosity.

Then, the elastic strain ε_1 responds to the unloading and the strain decreases, and then the plastic strain ε_3 of the combination of the spring and the dashpot represented by the Voigt model responds and the total strain decreases with the passage of time. Expressing these as equations, the total strain ε can be expressed by the following equation.

$$\begin{aligned} \varepsilon &= \varepsilon_1 + \varepsilon_2 + \varepsilon_3 \\ &= \frac{\sigma}{E_1} + \frac{\sigma}{\eta} t + \frac{\sigma}{E_2}\left\{1 - t\exp\left(\frac{\sigma}{\eta_2}\right)\right\} \end{aligned} \tag{5.7}$$

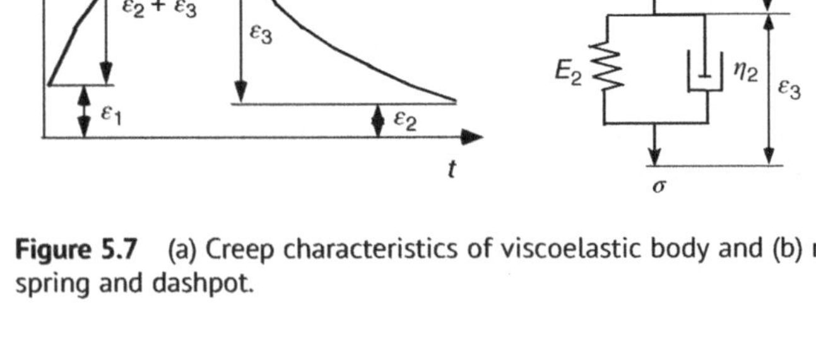

Figure 5.7 (a) Creep characteristics of viscoelastic body and (b) mechanical mode with spring and dashpot.

5.5 Conceptual Formula for Creep and Stress Relaxation

As mentioned above, there are two phenomena of stress relaxation and creep in the characteristic mechanical behavior of viscoelasticity. These phenomena will be explained based on the Maxwell model and Voigt model mentioned earlier.

5.5.1 Maxwell Model

The spring and dashpot are configured in series, and the relationship between the stress $\sigma(t)$ and the strain $\varepsilon(t)$ generated by this is from,

$$\sigma = \sigma_s = \sigma_d$$

$$\varepsilon = \varepsilon_s + \varepsilon_d$$

then,

$$\frac{d\varepsilon}{dt} = \frac{1}{k_s}\frac{d\sigma}{dt} + \frac{1}{\eta_d}\sigma$$

or,

$$k_s\frac{d\varepsilon}{dt} = \frac{d\sigma}{dt} + \frac{1}{\tau_M}\sigma \tag{5.8}$$

Here, $\tau_M = \frac{\eta_d}{k_s}$, note that τ_M is called relaxation time, which means that the stress decreases with the passage of time, and the rate of decrease is determined by the ratio of the viscosity η_d and the spring constant k_s.

5.5.2 Voigt Model

The spring and dashpot are configured in parallel, and the relationship between the stress σ_t and the strain ε_t generated by this is from,

$$\sigma = \sigma_s + \sigma_d$$

$$\varepsilon = \varepsilon_s = \varepsilon_d$$

then,

$$\boldsymbol{\sigma = \eta_d \frac{d\varepsilon}{dt} + k_s\varepsilon}$$

or,

$$\frac{d\varepsilon}{dt} + \frac{1}{\tau_V}\varepsilon = \frac{1}{\eta_d}\sigma \tag{5.9}$$

Here, $\tau_v = \frac{\eta_d}{k_s}$, which is called the delay time, means that the strain increases with the passage of time, but the rate of increase is determined by the ratio of the viscosity η_d and the spring constant k_s.

5.5.3 Conceptual Equation for Stress Relaxation and Creep

Using the two mechanical models explained earlier, the concept of the relationship between stress and strain of a general viscoelastic body is described. This is an indispensable mechanical model for deriving the constitutive equation of linear viscoelasticity theory in the range where the deformation is relatively small. First, as the initial conditions here, the differential equations for the Maxwell model and the Voigt model are shown under the conditions of stress $\sigma_t = 0$ and strain $\varepsilon_t = 0$ at time $t = -\infty$.

First, in the Maxwell model, the stress $\sigma(t)$ that changes over time is,

$$\frac{\mathrm{d}}{\mathrm{d}t}\sigma_t \exp\left(\frac{1}{\tau_M}\right) = E \exp\left(\frac{1}{\tau_M}\right)\frac{\mathrm{d}\varepsilon}{\mathrm{d}t} \tag{5.10}$$

Therefore, if this is integrated from $t = -\infty$ to $t = t$,

$$\sigma(t) = E\int_{-\infty}^{t} \exp\{-(t-\tau)/\tau_M\}\frac{\mathrm{d}\varepsilon(\tau)}{\mathrm{d}\tau}\mathrm{d}\tau \tag{5.11}$$

On the other hand, for the Voigt model, the following equation can be obtained in the same way.

$$\varepsilon(t) = \eta\int_{-\infty}^{t} \exp\{-(t-\tau)/\tau_M\}\frac{\mathrm{d}\sigma(\tau)}{\mathrm{d}\tau}\mathrm{d}\tau \tag{5.12}$$

Stress relaxation is a phenomenon in which stress σ decreases with the passage of time when an object is held under a constant strain ε_0. When this temporally changing stress is expressed by $\sigma(t)$,

$$E_r(t) = \frac{\sigma(t)}{\varepsilon_0}$$

is called relaxation modulus or stress relaxation function.

On the other hand, creep means that when a constant stress σ_0 is applied to an object, the strain is time.

A phenomenon that increases over time and changes in strain per unit stress.

$$D_c(t) = \frac{\varepsilon(t)}{\sigma_0}$$

is called creep compliance.

Now, the relaxation time τ_M defined above indicates the state of stress decrease. However, the details are left to technical books, and τ_M is the time when the stress becomes 37% of the initial stress. Therefore, the larger the value of τ_M, the slower the stress relaxation.

Now, as described above, the creep phenomenon and stress relaxation phenomenon of the viscoelastic body have been explained using the Maxwell model and the Voigt model, in addition, the conceptual contents are summarized in Figures 5.8 and 5.9.

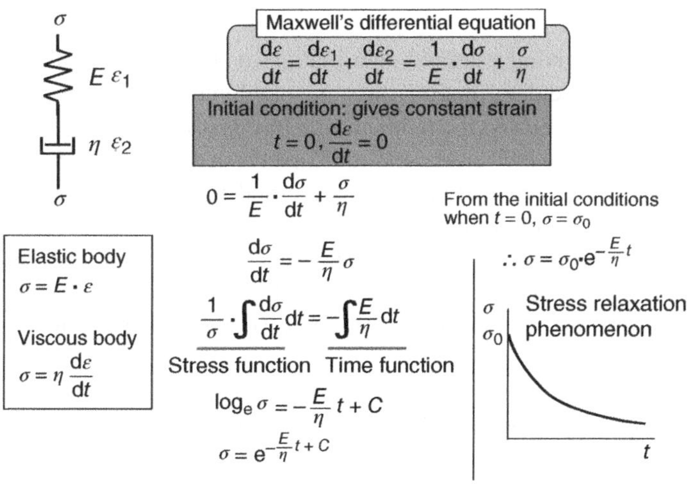

Figure 5.8 Mechanical model of viscoelastic body (Maxwell model).

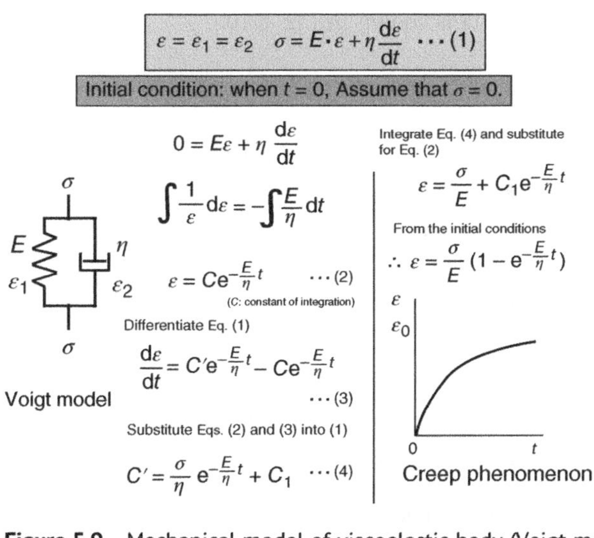

Figure 5.9 Mechanical model of viscoelastic body (Voigt model).

5.6 Master Curve and Time–Temperature Conversion Rule

In general, viscoelastic materials are greatly affected by time and temperature due to their mechanical and thermal properties. In addition, as mentioned above, there are two characteristic phenomena of creep and stress relaxation.

For example, in the stress relaxation phenomenon, a large stress is generated at the moment when a strain is applied to an object, and then decreases with the passage of time. This stress relaxation phenomenon changes greatly depending on the temperature.

Now consider the case of giving a desired deformation to an object. If it takes a long time to obtain the desired deformation, the same phenomenon can generally be achieved in a short time by raising the temperature. In this way, when the same phenomenon is obtained, there is a certain relationship between time and temperature. The time–temperature conversion law is a method for obtaining such a stress relaxation phenomenon on a time scale.

For example, in the Maxwell model that expresses viscoelastic behavior, the elastic modulus of a spring is not affected by temperature, but the viscoelastic coefficient is greatly affected by temperature and decreases significantly as the temperature rises.

Figure 5.10 shows an example of actually measuring the elastic modulus of a resin in the temperature range of 50–200°C at different frequencies, and the master curve is explained based on this.

In Fig. 5.10a, the elastic modulus is shown on the vertical axis and the reciprocal of the frequency, that is, the actual time (real time), on the horizontal axis, and the elastic modulus for each measured temperature is displayed. Using the time–temperature conversion rule, this data is shown in Fig. 5.10b.

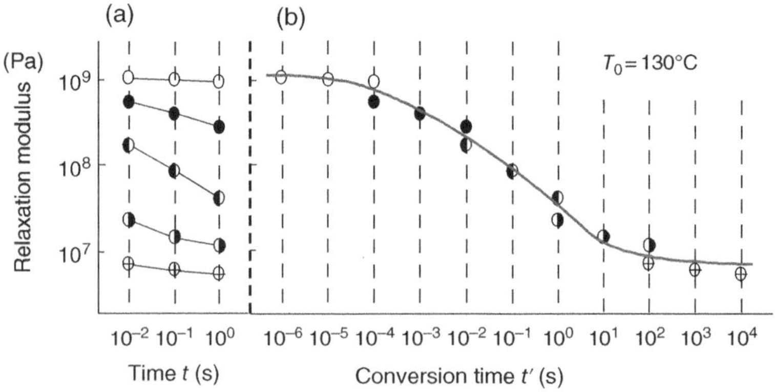

Figure 5.10 Creating a master curve. (a) Real time and (b) master curve with converted time.

If the measured value at low temperature is shifted to the left and the measured value at high temperature is shifted to the right with the reference temperature T_0 as 130°C. By plotting on the conversion time axis as shown earlier, one smooth master curve can be obtained by overlapping each other.

Here, the amount of shift in which the measured data in real time is translated to the right or left on the conversion time axis is called the time–temperature shift factor $a_{T0}(T)$. That is, $a_{T0}(T)$ can be regarded as a kind of acceleration rate because the time can be shortened by increasing the temperature.

5.7 Approximation of Master Curve

Figure 5.11 briefly explains how to approximate the master curve of the elastic modulus of the resin with the Prony series. This is substituted into the calculation formula of viscoelasticity in the numerical analysis of viscoelastic stress/deformation described later.

That is, the relaxation elastic modulus E_r and the master curve represented by the conversion time t' are approximated by the Prony series of the following equation.

$$E_r(t') = E_0 + \sum_{i=1}^{n} E_i \exp\left(\frac{t'}{\tau_i}\right) \tag{5.13}$$

Here, the relaxation time $\tau_i = \dfrac{\eta}{E}$

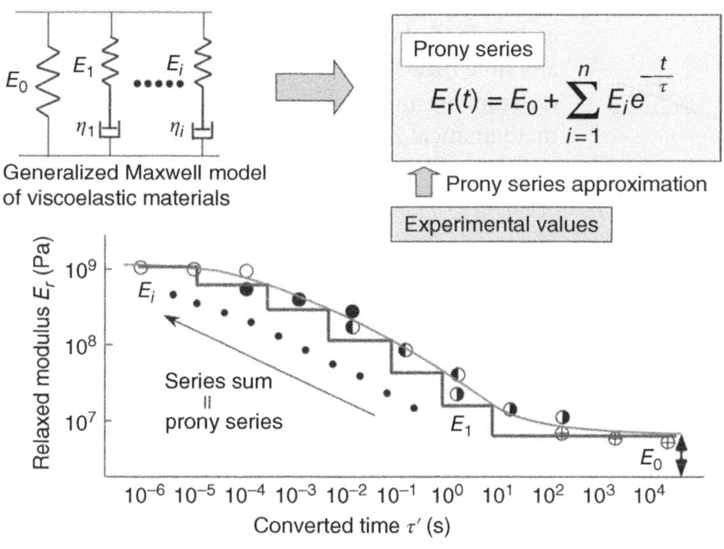

Figure 5.11 Approximation of Prony series and master curve.

In this example, the elastic modulus is E_0 when the conversion time t' is 10^4 for a long time, and it is a series sum showing a tendency that the elastic modulus increases as the conversion time becomes shorter.

For reference, Fig. 5.11 shows a generalized Maxwell model that models the Prony series. Actual materials such as resins cannot express their viscoelastic behavior with such a simple model, but they are useful for understanding the concept of viscoelastic behavior.

5.8 Superposition Principle and Basic Equations

When a load is applied to an object, a response such as strain or deformation occurs to it. The linearity between the strain ε and the stress σ means that the stress $\sigma(\varepsilon_1)$ with respect to the strain ε_1 and the stress $\sigma(\varepsilon_2)$ with respect to the strain ε_2 generally mean that the stress $\sigma(\varepsilon_1 + \varepsilon_2)$ with respect to $(\varepsilon_1 + \varepsilon_2)$ is expressed by equation (5.14), and there is a proportional relationship between stress and strain.

$$\sigma(\varepsilon_1 + \varepsilon_2) = \sigma(\varepsilon_1) + \sigma(\varepsilon_2) \tag{5.14}$$

It can be expressed by, and a proportional relationship is established between the two.

Here, if a new strain is applied to the strain-applied system of the object at time t_i, the change in stress after that is the stress caused only by the strain applied at t_i at the beginning. That is, when repeated strains are applied, the stress for each strain is independent, and this is called the superposition principle and is explained in Fig. 5.12.

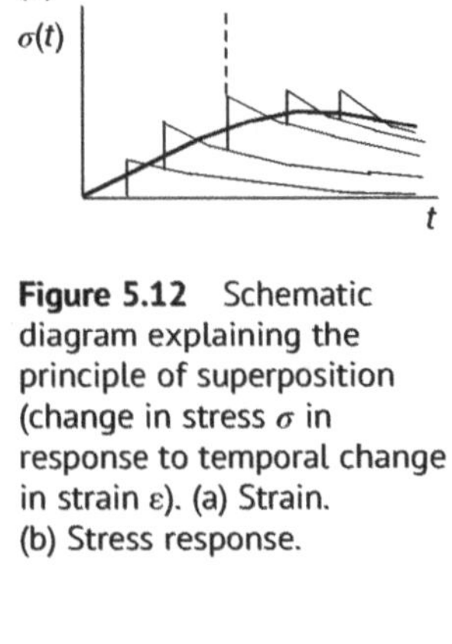

Figure 5.12 Schematic diagram explaining the principle of superposition (change in stress σ in response to temporal change in strain ε). (a) Strain. (b) Stress response.

First, when the applied strain changes with time, it can be divided into the stepwise strain $\Delta\varepsilon_i$ applied to any time t_i as shown in Fig. 5.12a, so that the temporal response of stress $\sigma(t)$ can be expressed by the mathematical relation of the following equation.

$$\sigma(t) = \sum \Phi(t - t_i)\Delta\varepsilon_i(t_i) \tag{5.15}$$

Here, $\varphi(t - t_i) \cdot \Delta\varepsilon_i(t_i)$ is the value of the stress caused by the strain $\Delta\varepsilon_i$ at time t_i at time t. This means that the response to the stimulus is uniquely determined as a function of $(t - t_i)$. That is, the stress can be determined by summing up all the processes of the applied strain, which is schematically shown in Fig. 5.12b.

Furthermore, considering the case where the strain is continuously changed from the reference state ($\varepsilon = 0$) in the distant past ($t' = -\infty$) by making the division interval of t sufficiently small, the stress

$\sigma(t)$ that changes with time can be expressed by the integral form of the following equation.

$$\sigma(t) = \int_{-\infty}^{t} \varphi(t-t') \frac{d\varepsilon(t')}{dt'} dt' \tag{5.16}$$

This equation is called the basic equation of Boltzmann. Here, the function $\varphi(t-t')$ is a characteristic function that determines the nature of the response of the system to the stimulus and is called the relaxation function or the aftereffect function.

5.9 Simple Model of Generating Thermal Stress and Strain

Using the methods and processes described earlier, I have developed Visco-Elastic Stress Analysis Program (VESAP) software that uses coupled analysis of thermal and stress analysis to determine the thermal stress, strain, and warpage that occur when a thermal load acts on an LSI package.

A simple mechanism concept of how thermal stress or residual stress is generated is shown in Fig. 5.13 using resin and an LSI chip.

From Fig. 5.13, it can be seen that heat causes expansion and contraction of both materials, and that these changes can be represented by springs and dashpots.

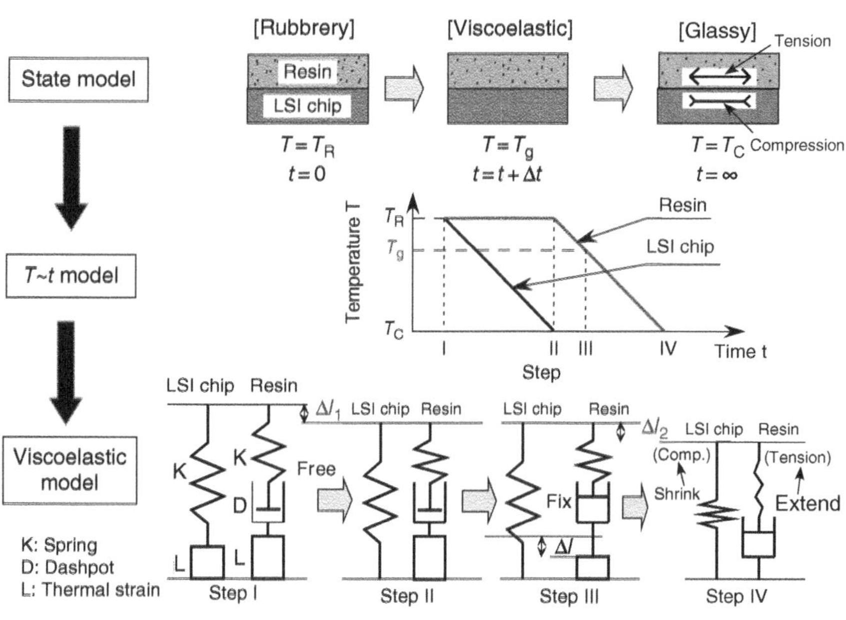

Figure 5.13 Concept of thermal stress generation mechanism.

6

Measurement of Viscoelastic Properties

6.1 Dynamic Viscoelasticity

In this chapter, I first describe dynamic viscoelasticity. Viscoelasticity is the property that when a constant strain ε_0 is applied to an object, stress $\sigma(t)$ is generated in the object, and the generated stress $\sigma(t)$ decreases with the passage of time.

Here, if the relationship between the generated stress $\sigma(t)$ and the applied strain ε_0 is the relaxation modulus $E(t)$, $E(t)$ is expressed as follows:

$$E(t) = \frac{\sigma(t)}{\varepsilon_0} \tag{6.1}$$

In this way, the static measurement method is to investigate the stress response to a static strain load. This static measurement method is suitable for measuring viscoelasticity on a simple and long-time scale.

However, with materials such as rubber and chewing gum, viscoelasticity cannot be evaluated by the static measurement method for dynamic loads (frequency-dependent). The dynamic measurement method measures the temporal response of stress and strain to such a dynamic load, and the property of viscoelasticity measured by that method is dynamic viscoelasticity.

The actual measurement of dynamic viscoelasticity is to measure the stress response that changes over time when the test piece is subjected to forced vibration of sinusoidal strain. The details are described here.

6.2 Measurement Method

To measure the viscoelastic properties of a viscoelastic material such as resin, chuck the specimen in the chamber of the measuring device as shown in Fig. 6.1.

Polymeric Materials for Electronic Packaging, First Edition. Shozo Nakamura.

Published 2023 by John Wiley & Sons, Inc.

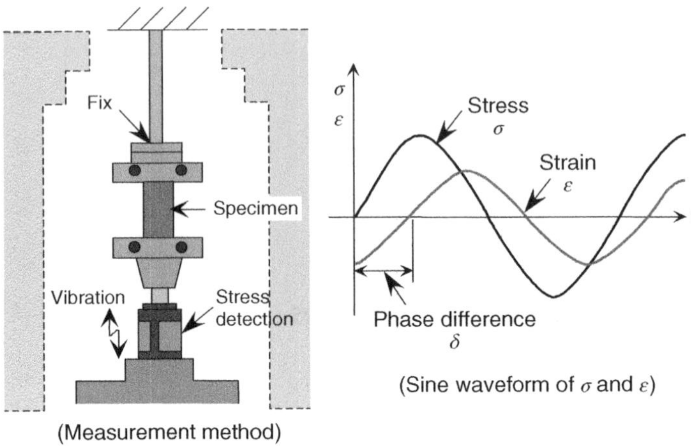

Figure 6.1 Dynamic viscoelastic measurement principle.

The relationship between the strain ε and the time t at the c-terminus can be expressed by the following equation, where ω is the angular frequency.

$$\varepsilon = \varepsilon_0 e^{i\omega t} \tag{6.2}$$

Alternatively, if the strain rate $d\varepsilon/dt$ is used instead of strain,

$$\frac{d\varepsilon}{dt} = i\omega\varepsilon \tag{6.3}$$

The stress σ generated at the end by such forced vibration is a sinusoidal vibration in the steady state, but the phase of the stress σ does not always match the phase of the strain ε.

The reason is that, as a characteristic of the test piece, the deformation is generally viscous, which causes some time lag. However, if the test piece is a completely elastic body and its deformation exhibits completely elastic behavior, no phase difference will occur.

Therefore, this phase shift is greatly related to the elasticity and viscosity of the test piece.

6.3 Complex Modulus and Mechanical Model

The elasticity and viscosity of the test piece when the test piece is forcibly vibrated at the angular frequency ω (or frequency $f = \omega/2\pi$) are displayed by the viscoelastic dynamics model of the spring and dashpot shown in Fig. 6.2.

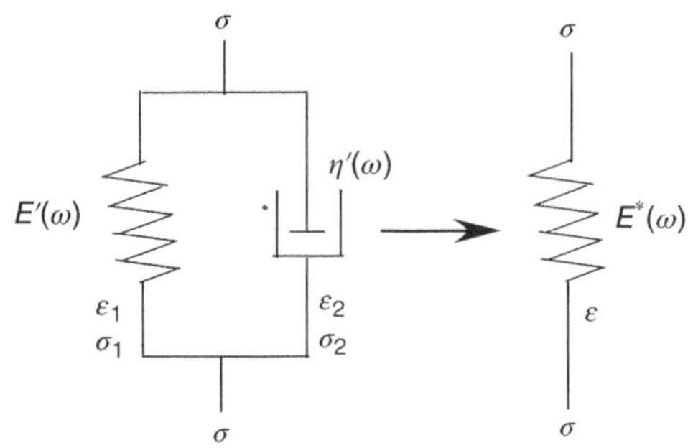

Figure 6.2 Viscoelastic dynamics model.

Here, let the spring constant be $E'(\omega)$ and the viscosity coefficient (viscosity) of the dashpot be $\eta'(\omega)$. In viscoelastic analysis, $E'(\omega)$ and $\eta'(\omega)$ are called dynamic modulus and dynamic viscosity, respectively.

Next, this viscoelastic dynamics model is modeled with a single spring as shown on the right in Fig. 6.2, and since this spring constant is represented by a complex number, it is called the complex modulus $E^*(\omega)$ and is expressed by the following equation.

$$E^*(\omega) = \frac{\sigma}{\varepsilon} \tag{6.4}$$

Here, in the range of linear viscoelasticity where the deformation of the test piece is small, Hooke's law holds for the spring.

$$\sigma_1 = E'\varepsilon_1 \tag{6.5}$$

As for the dashpot, Newton's law of viscosity holds.

$$\sigma_2 = \eta' \frac{d\varepsilon_2}{dt} \tag{6.6}$$

Note that σ_1 and σ_2 are the stresses shared by the spring and the dashpot, respectively. That is, the total stress is

$$\sigma = \sigma_1 + \sigma_2 \tag{6.7}$$

From this equation, the complex elastic modulus $E^*(\omega)$ can be expressed by the following equation.

$$\begin{aligned} E^*(\omega) &= \frac{\sigma}{\varepsilon} = E'(\omega) + i\omega\eta' \\ &= E'(\omega) + iE''(\omega) \end{aligned} \tag{6.8}$$

Here, the imaginary part $E''(\omega) = \omega\eta'$ is called dynamic loss. In other words, E' is the component of the energy generated by external force and strain on the object that is stored inside the object, while E'' is the component that diffuses to the

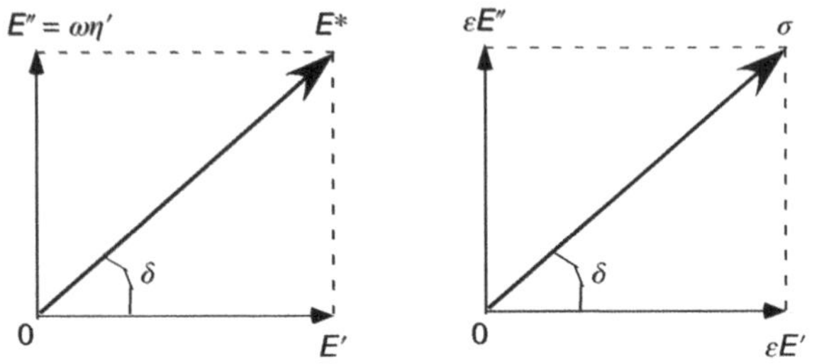

Figure 6.3 Vector display of viscoelastic properties.

outside. Simply, E'' is a component that indicates the degree of viscosity of an object, and a large E'' means that the viscosity is high, as shown in Eq. (6.9).

It affects the value of $\tan \delta$, and this value serves as a guide for the temperature of softening of the object (glass transition temperature). It is an important component that is indispensable for the selection of materials used in actual electronic equipment and the simulation when calculating thermal stress and thermal deformation.

Figure 6.3 shows Eq. (6.8) as a vector diagram. Here, as a convention, if the phase difference angle is δ,

$$\tan \delta = \frac{E''}{E'} \tag{6.9}$$

There is a relationship. When the viscosity $\eta' = 0$, that is, the completely elastic behavior of $E'' = 0$ is exhibited, $\delta = 0$, and the stress σ and the strain ε are in phase.

6.4 Dispersion and Absorption by Frequency

Consider a material that can express viscoelasticity with multiple Maxwell models shown in Fig. 6.4, which is ideally simplified. Then, let us consider how the dynamic elastic modulus E' and the dynamic loss E'' change depending on the angular frequency ω in this viscoelastic dynamics model.

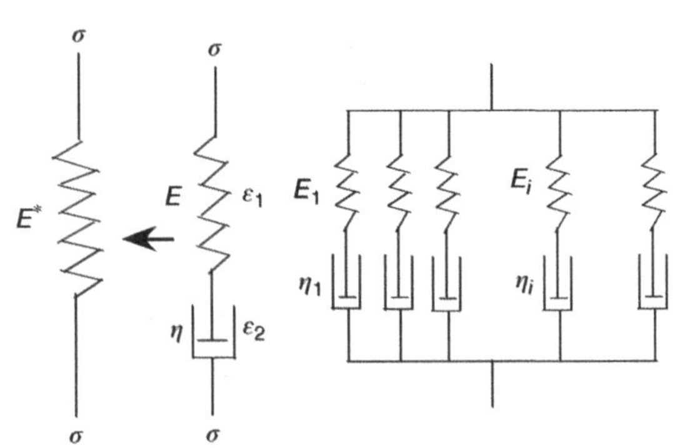

Figure 6.4 Modeling of multiple Maxwell models.

That is, when an external force is applied to an object, stress σ and strain ε are generated, which differ depending on the velocity of the external force (angular frequency ω), and as a result, affect the dynamic elastic modulus of the object.

This dynamic modulus is an important parameter indispensable for viscoelasticity simulation at the time of mounting.

Calculate the complex dynamic modulus E^* from the stress σ and the strain ε.

In the Maxwell model, $\boldsymbol{\varepsilon = \varepsilon_1 + \varepsilon_2, \varepsilon_1 = \frac{\sigma}{\varepsilon}, \varepsilon_2 = \frac{\varepsilon}{i\omega} = \frac{\sigma}{i\omega\eta}}$,

$$\begin{aligned} E^*(\omega) &= \frac{\sigma}{\varepsilon} = \frac{\sigma}{\omega_1 + \omega_2} \\ &= \frac{\omega^2\tau^2 E}{1+\omega^2\tau^2} + \frac{\omega\tau E}{1+\omega^2\tau^2} i \end{aligned} \tag{6.10}$$

Here, with $\tau = \eta/E$, τ is the relaxation time that characterizes viscoelastic behavior, and is an index that expresses the properties of elasticity and viscosity of an object. Simply put, the object changes from a hard state to a soft state. It can be a guide for temperature and time.

The dynamic elastic modulus E' and the dynamic loss E'' can be expressed by the following equations.

$$E' = \frac{\omega^2\tau^2 E}{1+\omega^2\tau^2} \tag{6.11}$$

$$E'' = \frac{\omega\tau E}{1+\omega^2\tau^2} \tag{6.12}$$

Figure 6.5 shows the changes in E' and E'' using $\log \omega t$ on the horizontal axis instead of ω and E on the vertical axis. τ is a constant and the smaller the value on the horizontal axis, the higher the frequency.

From Fig. 6.5, the dynamic elastic modulus E' decreases significantly as the frequency decreases. This phenomenon is called the frequency dispersion of elastic modulus, and E'' shows a maximum value at $\omega t = 1$, which is called the maximum absorption because it is the maximum of energy absorption.

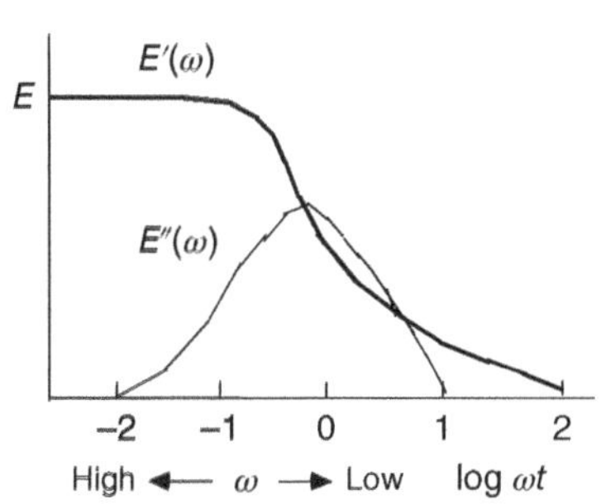

Figure 6.5 Relationship between frequency, dynamic modulus, and dynamic loss.

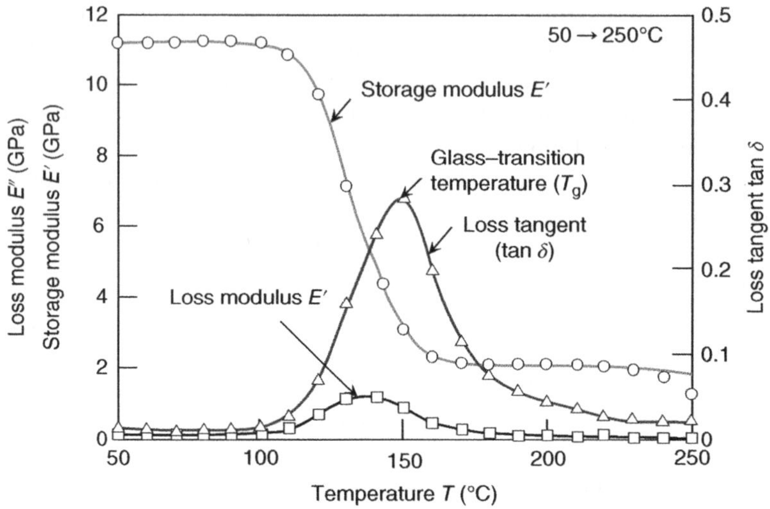

Figure 6.6 Measurement of temperature and viscoelastic properties (sample of FR-4 substrate).

6.5 Actual Measurement Example

To measure the viscoelastic properties of a viscoelastic material such as resin, chuck the specimen in the chamber of the measuring device as shown in Fig. 6.1.

At this time, the temperature inside the chamber is changed and the vibration frequency is set to a desired value. Figure 6.6 shows an example of the results measured by such a method.

Figure 6.6 shows the case where the frequency is constant at 10 Hz. From Fig. 6.6, the storage elastic modulus E', the loss elastic modulus E'', and the loss tangent $\tan\delta$ are output in the process of increasing the measured temperature.

7

Design Issues of LSI Packages

7.1 Introduction

As mentioned in Chapter 1, electronic devices are represented by smartphones and iPads and are becoming lighter, thinner, shorter, and smaller as well as have higher functionality and performance.

Along with this, semiconductor packages and printed circuit boards have become diversified, and their mounting technology has undergone remarkable changes and progress. Large-scale integrated (LSI) packages and printed circuit boards are made of various different materials such as resins, and in the manufacturing process, heat load due to high-temperature solder reflow of about 250°C or higher and various reliability tests after that are imposed.

Due to the heat load, large thermal stress and warpage deformation occur in semiconductor packages due to differences in the physical properties of the constituent materials, which cause cracks in the sealing resin, peeling of the adhesive interface, and even large warp deformation, resulting in large warp deformation. There are also many technical issues in mounting design.

In recent years, there has been a movement in large-area batch mounting in search of manufacturing cost reduction and efficient mounting technology in each company. For this reason, the technology for mounting on semiconductor packages and printed circuit boards has become a strict requirement, and the technical problems have become complicated and difficult.

In order to solve this kind of problem, and to design highly reliable electronic devices such as semiconductor packages and printed circuit boards, it is essential to design the resin, heat load conditions in the manufacturing process, and optimize the structure and shape. In particular, unlike metals and ceramics, resins exhibit so called viscoelastic properties, which are greatly affected by temperature and time, so they are a major obstacle in the optimal design of electronic devices that incorporate a large number of LSI packages.

Polymeric Materials for Electronic Packaging, First Edition. Shozo Nakamura.

Published 2023 by John Wiley & Sons, Inc.

Therefore, in the manufacture of highly reliable electronic equipment, it is even more important to construct an appropriate stress and deformation simulation technique that fully considers the viscoelastic properties of the resins.

In order to solve the technical problems of such a thin and small LSI package, the author has developed a simple and convenient viscoelastic analysis software Visco-elastic Stress Analysis Program (VESAP), which is described in detail in Chapter 6.

I will demonstrate that VESAP analysis is a useful tool for optimal design of LSI packages. In addition, Fig. 7.1 conceptually summarizes the necessity of analysis considering the viscoelastic properties of the resin material in the optimum design of LSI package.

In this chapter, from the above background, a specific LSI package is taken as an example, and its trends and technical issues are described.

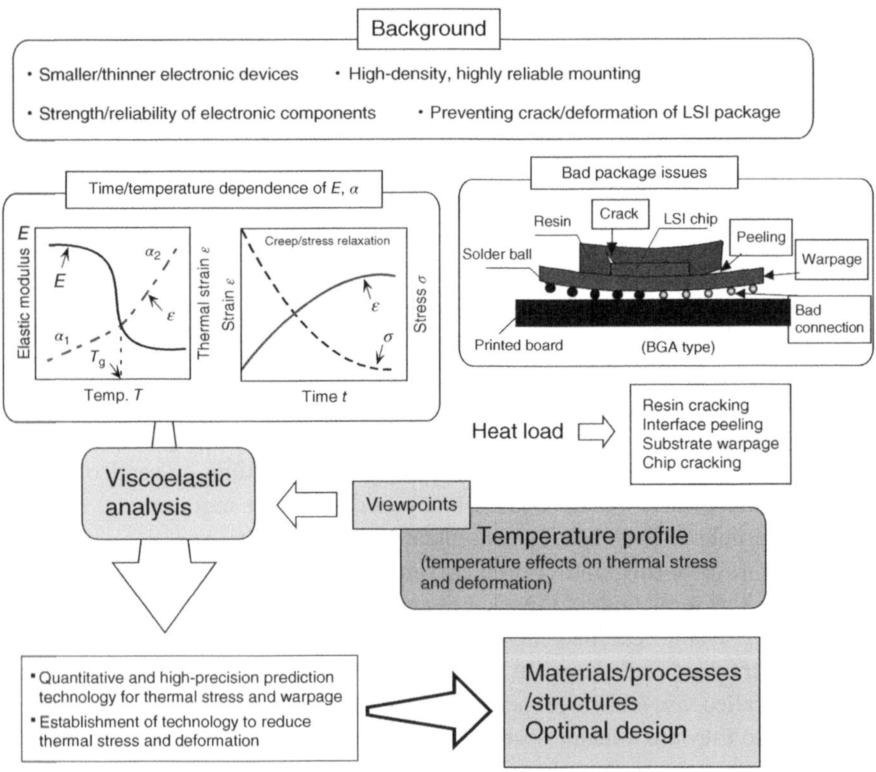

Figure 7.1 Research purpose and background.

7.2 Trends and Issues of LSI Packages

As mentioned above, the structure and shape of semiconductor packages have changed significantly, and the structures and shapes also vary from company to company. The ultimate in semiconductor packages and board mounting used in various electronic devices is the chip scale package (CSP), which is reduced to a chip size, and CSPs are currently the mainstream and are widely used.

Semiconductor packages up to CSP generally have a structure in which an LSI chip is sealed with an epoxy resin filled with a filler.

Here, as an example, various problems will be described by taking as an example a structure in which a semiconductor package having an over molded pad array carrier (OMPAC) package structure shown in Fig. 7.2 is mounted on a printed circuit board.

The general process of the mounting process is as follows.

① Mount the LSI chip on the sub-board.
② Wire bond with the desired Au wire.
③ Seal the LSI chip with resin by truss fur molding, etc.
④ Attach the solder ball to the sub-board.
⑤ This semiconductor package is bonded to a printed circuit board at a temperature of about 250°C by solder reflow.

The following technical issues are expected in the manufacturing process that goes through such various processes. First, the thermal stress generated by the

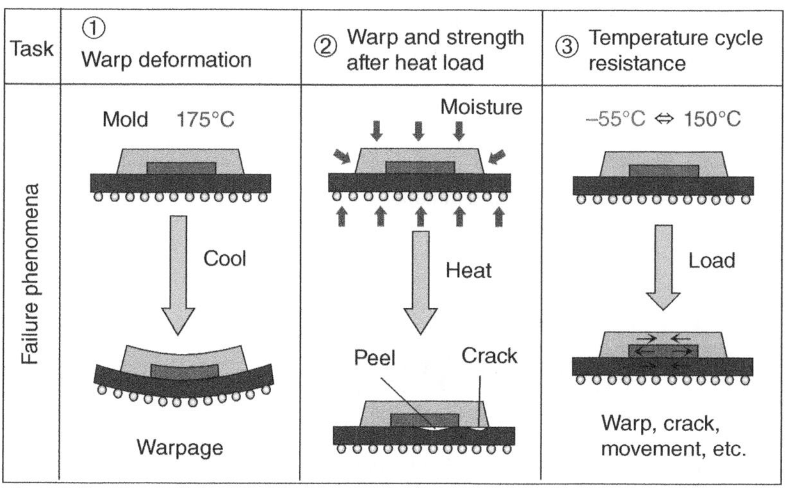

Figure 7.2 Failure example of OMPAC type LSI package.

temperature of about 250°C loaded in the solder reflow process causes cracks in the sealing resin and peeling of the adhesive bonding interface.

In addition, since semiconductor packages are generally composed of different materials, the difference in the amount of thermal expansion caused by the difference in the coefficient of linear expansion causes warpage deformation of the semiconductor package, printed circuit board, etc., which is shown in Fig. 7.2. There is a possibility that such poor adhesion between the solder ball and the printed circuit board may occur.

As shown in Fig. 1.1a, b in Chapter 1, there are a wide variety of semiconductor packages, and any package can be taken up.

In recent years, semiconductor packages are small and the BGA type is often used, and they are composed of various materials with different properties such as substrates, encapsulants, and LSI chips. It was taken up as a representative example because it is easy to understand when explaining the problems and issues of semiconductor package.

From upper reasons, here, an example of an OMPAC type ball grid array (BGA) package is given in Fig. 7.3, but in the case of a semiconductor package with a combination of thin materials, a large-area printed circuit board or Si interface, or a CSP with a different structure and shape, etc.

Due to the heat load in the manufacturing process, problems such as warpage deformation and adhesive-bonded interface peeling become more prominent. For semiconductor packages with various forms to be developed in the future, resin materials can be used to manufacture highly reliable and high-performance semiconductor packages that do not cause defects due to heat load during the manufacturing process and reliability test. In addition to optimizing the structure and shape, as well as the heat load conditions, elucidating the mechanism of defect occurrence and establishing drastic countermeasures will be an even more important technical issue.

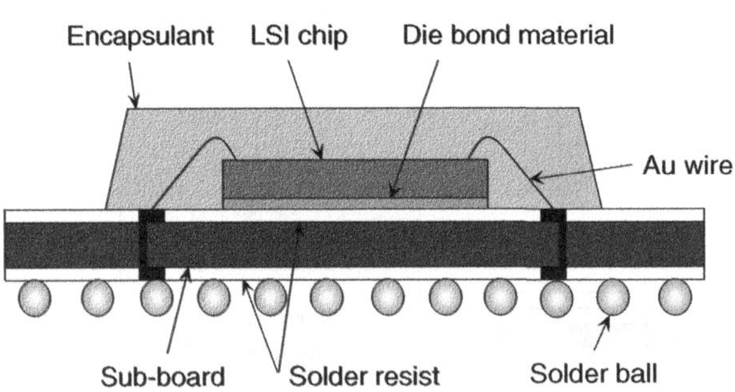

Figure 7.3 Sectional structure of OMPAC type BGA package.

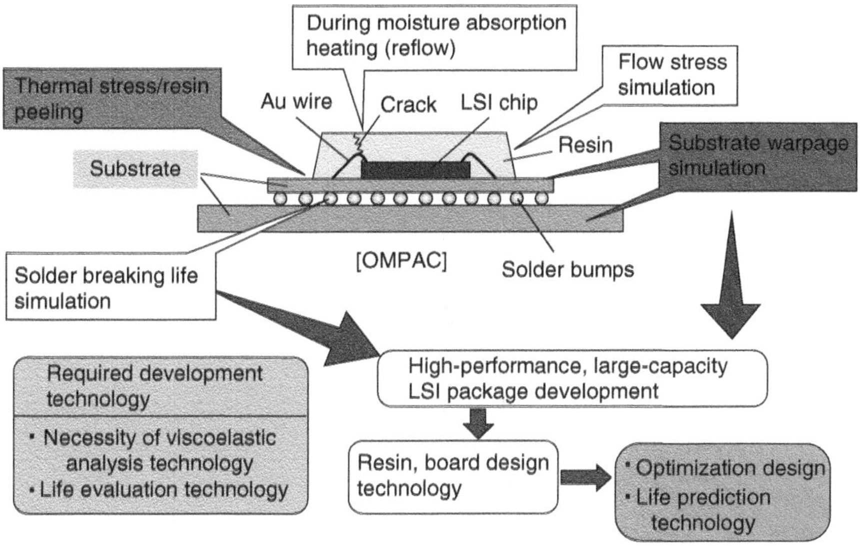

Figure 7.4 Outline of future LSI package development technology.

Figure 7.4 summarizes the outline of the LSI package development technology that will be required in the future. Among these are the development of various simulation technologies such as flow analysis technology and life prediction technology that take resin viscosity into consideration, and life prediction, which are necessary over the long term.

8

Validity of Viscoelastic Analysis

8.1 Introduction

I mentioned that I developed viscoelastic analysis software – Visco-Elastic Stress Analysis Program (VESAP) – described in the appendix chapter. In this chapter, the validity of the VESAP analysis is verified for various laminates assuming LSI packages.

In other words, the validity and effectiveness of the VESAP analysis are examined by comparing the numerical values obtained by the VESAP analysis with the experimental values for the complex multi-layer model from the simple two-layer laminate model.

It also mentions the usefulness of viscoelastic analysis by comparing it with the results of elastic analysis by the finite element method (FEM) based on the conventional elasticity theory.

The specific laminated body model is as follows. That is, as the most basic model of multi-layer laminate

1) Two-layer laminate consisting of Epoxy resin/Al alloy
2) Two-layer laminate consisting of an actual encapsulating resin for LSI packages/Fe–Ni alloys (42 alloys).

The author will describe an example of comparing the VESAP analysis with the actual measured values and the elastic analysis by the FEM.

Specifically, for (1), two types of cooling methods, slow cooling and rapid cooling, from the holding temperature T_h of the epoxy resin glass transition temperature T_g or higher to the cooling temperature T_c are taken up, and for (2), similarly, the case of cooling from a high temperature was taken up, and the amount of warpage deformation was compared between the VESAP analysis considering the viscoelastic properties of the epoxy resin, the elastic analysis by FEM, and the measured values. Figure 8.1 shows the procedure of validation for VESAP analysis.

Polymeric Materials for Electronic Packaging, First Edition. Shozo Nakamura.

Published 2023 by John Wiley & Sons, Inc.

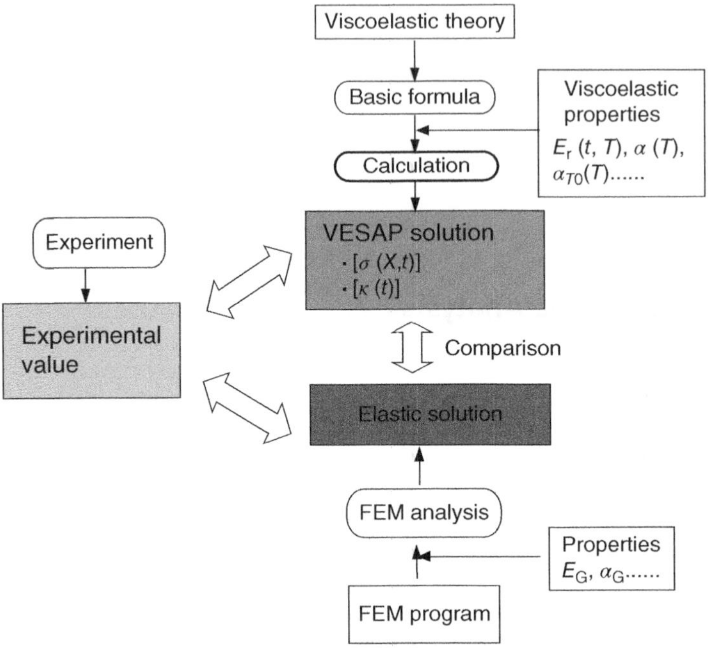

Figure 8.1 Comparison of VESAP solution with experimental value and FEM solution.

8.2 Structure of Laminated Body

8.2.1 In the Case of a Laminate of Epoxy Resin/Al Alloy

A laminate consisting of an epoxy resin and an elastic Al alloy shows in Fig. 8.2. Table 8.1 shows the thermal and mechanical properties of these constituent materials. Epoxy resin shows linear viscoelastic behavior at a strain of 1.5% or less in the

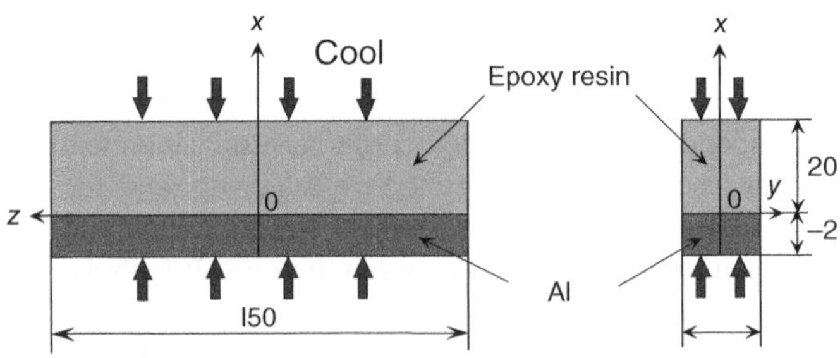

Figure 8.2 Shape and coordinates of two-layer laminate.

Table 8.1 Physical characteristics of materials.

Property / Material	Temperature conductivity (m^2/s)	Glass-transition temperature T_g (°C)	Coefficient of linear expansion (10^{-6}/K)		Relaxation modulus (GPa)	
			$T \leqq T_g$	$T < T_g$	$T \leqq T_g$	$T < T_g$
Epoxy resin	1.09×10^{-7}	89	88	217	3.4	0.03
Al alloy	9.22×10^{-5}	—	24		67	

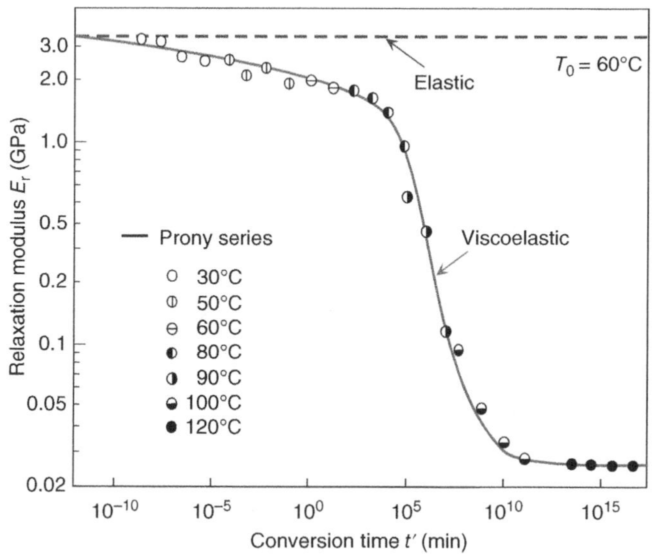

Figure 8.3 Relaxation modulus and master curve of epoxy resin.

temperature range from room temperature to 180°C, and the master curve of its relaxation modulus is shown in Fig. 8.3, and the time–temperature shift factor at this time is shown in Fig. 8.4.

Here, the dimensions of the laminate were set so that the depth $a = 2$ mm in the Al alloy portion and the depth $b = 20$ mm in the epoxy resin portion (Fig. 8.2).

8.2.2 Laminate of Encapsulating Resin/Fe–Ni Alloy (42 Alloy)

The laminate was composed of an LSI encapsulating resin and a Fe–Ni alloy (42 alloy) for a lead frame. Here, there are two types of resin for encapsulating LSI, resin A and biphenyl-type resin B.

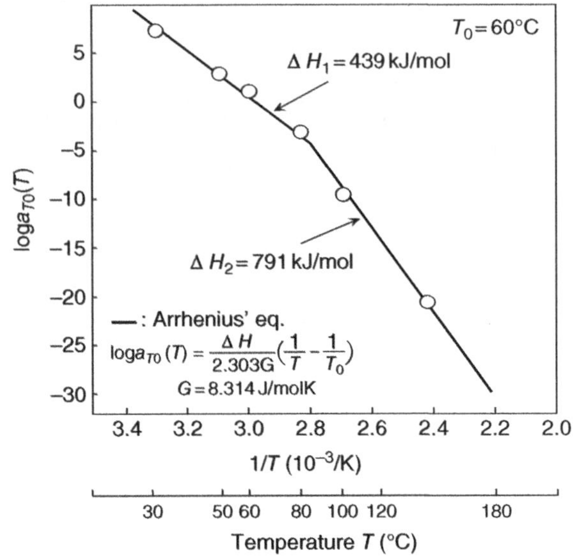

Figure 8.4 Time-temperature shift factor.

The material properties of Resin A, Resin B, and 42 Alloy are shown in Table 8.2. These resins are made by filling the main agent epoxy resin, with 60–70 vol% of molten quartz filler.

Figure 8.5 shows the shape of laminate, and Figs. 8.6 and 8.7 show the master curves of the relaxation modulus of resin A and resin B, respectively.

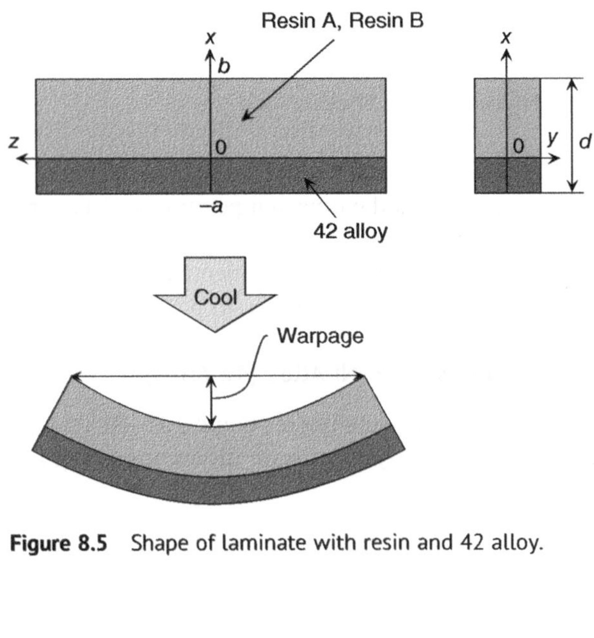

Figure 8.5 Shape of laminate with resin and 42 alloy.

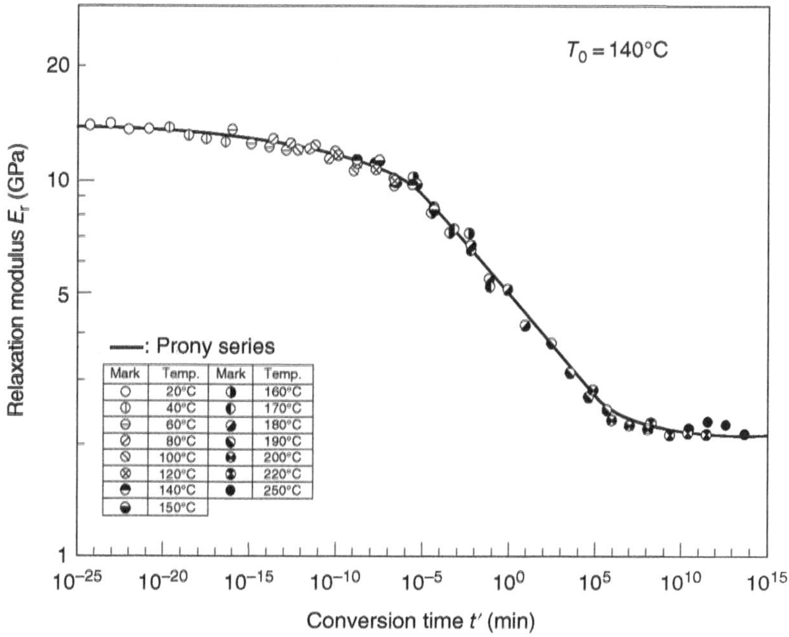

Figure 8.6 Master curve of relaxation modulus of resin A.

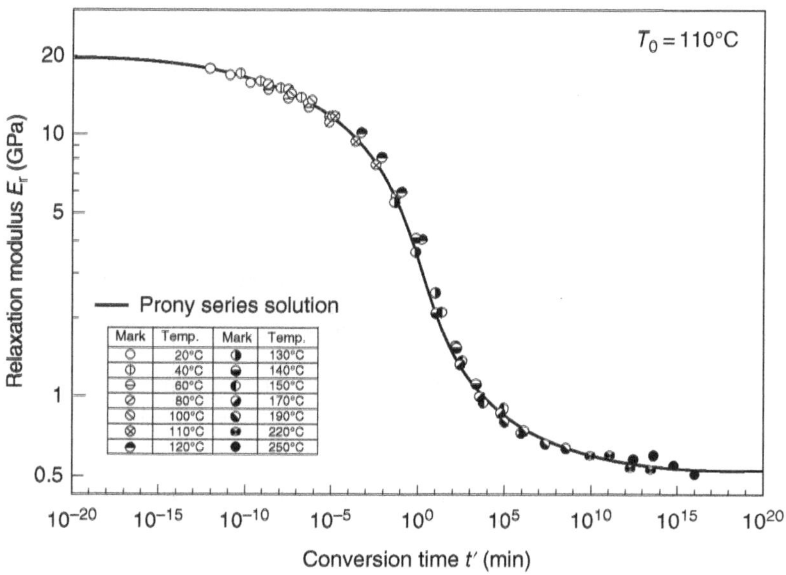

Figure 8.7 Master curve of relaxation modulus of resin B.

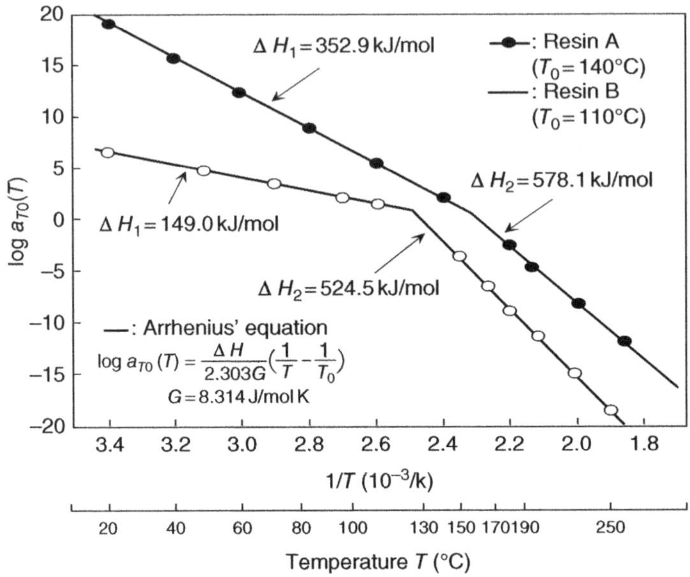

Figure 8.8 Time–temperature shift factor of resin A and resin B.

From Figs. 8.6 and 8.7, one smooth master curve was obtained for both resins from a very short time to a very long time. In addition, as shown in Fig. 8.8, the time–temperature shift factor at this time can be expressed by two Arrhenius equations in which the activation energy ΔH differs between the low-temperature part and the high-temperature part.

In this way, both resins A and B show remarkable viscoelastic behavior, and the time–temperature conversion rule is established.

Table 8.2 Material property of laminate.

Property / Material	Thermal conductivity (W/m K)	Specific heat (kJ/kg K)	Density (kJ/m^3)	Glass-transition temperature T_g (°C)	Expansion coefficient (10^{-6}/K)		Elastic modulus (GPa)	
					$T \leqq T_g$	$T < T_g$	$T \leqq T_g$	$T < T_g$
Resin A	0.753	0.950	1870	152	16.4	65	15.8	1.2
Resin B	0.879	0.950	1870	113	13.1	48	20.0	0.3
42 alloy	15.93	0.460	8120	—	5		150	

Table 8.2 shows the physical property values of other constituent materials. A supercomputer manufactured by Hitachi, Ltd. was used for the VESAP analysis of the warp deformation.

8.3 Analysis Method

In the VESAP Analysis, the physical properties of Tables 8.1 and 8.2 and the master curves of Figs. 8.3, 8.6, and 8.7 are approximated by Prony series, and the time–temperature shift factors in Figs. 8.4 and 8.8 are shown in the high-temperature part and low-temperature part.

Approximate with two Arrhenius equations with different activation energies ΔH, and these values were used. Regarding the temperature condition at the time of analysis, regarding (1), after holding the laminate shown in Fig. 7.2 at the holding temperature $T_h = 180°C$.

i) When slowly cooling to room temperature $T_e = 20°C$ at a cooling rate of about 1°C/min
ii) When quenching in water to 20°C

The two conditions were set. The temperature conductivity in Table 8.1 was used to calculate the temperature distribution inside the laminate in the case of quenching.

For the VESAP analysis, the above equation was expressed in a difference format and performed by the sequential imputation method.

In addition, the above (2) was also analyzed by basically the same method as (1).

On the other hand, in the elastic analysis by general-purpose FEM, the values of the room temperature epoxy resin, Al alloy, sealing resin, 42 alloy's Young's modulus, and linear expansion coefficient shown in Tables 8.1 and 8.2.

These analysis results will be described later in comparison with the experimental results.

8.4 Cooling Experiment of Laminated Body

The laminate used for the VESAP analysis was actually manufactured, slow cooling and quenching experiments were performed under the same temperature conditions as the analysis, and the amount of warpage deformation during cooling and the residual stress distribution after cooling were measured by the sequential removal method.

8.4.1 Test Piece

Epoxy resin strips and Al alloy strips were prepared and adhesively bonded in an atmosphere of 180°C to prepare the laminate shown in Fig. 8.2.

8.4.2 Cooling Method

In the slow cooling experiment, the laminate was sufficiently maintained in a heating atmosphere of 180°C and then cooled to room temperature of 20°C at a cooling rate of 1°C/min.

On the other hand, in the quenching experiment, transparent silicone rubber plates, which are heat-insulating materials, were attached to both sides of the laminate and held sufficiently at a holding temperature of 180°C. It was kept in running water until the temperatures became almost equal.

While confirming the temperature uniformity in the slow cooling experiment by the thermocouple embedded in the laminate, it was also confirmed that the temperature in the quenching experiment was distributed only in the depth direction as shown in Fig. 8.9.

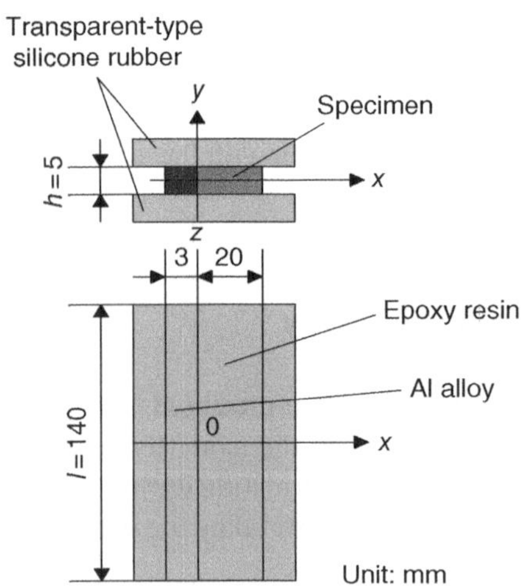

Figure 8.9 State of the test piece of quenching.

8.5 Analysis Results and Experimental Values

8.5.1 In the Case of a Laminate of Epoxy Resin/Al Alloy

Regarding the deformation during cooling, the case of slow cooling is omitted, and the case of rapid cooling is shown. The upper part of Fig. 8.10 shows the temporal changes on the surface and inside of the laminate when the laminate is rapidly cooled.

The theoretical solution is obtained from the equation of unsteady heat conduction assuming that the cooling temperature of both surfaces of the laminate instantly reaches 20°C at $t = 0$, but the calculated temperature distribution that changes with time is It was in good agreement with the value measured by the thermocouple inside the laminate.

Figures 8.11 and 8.12 show the VESAP solution, elastic solution, and experimental values of the expansion, contraction, and curvature that change from moment to moment when the laminated body is rapidly cooled.

Although there is almost no difference in expansion and contraction between the three solutions, the VESAP solution of curvature and the elastic solution are significantly different, and the experimental values are in excellent agreement with the VESAP solution.

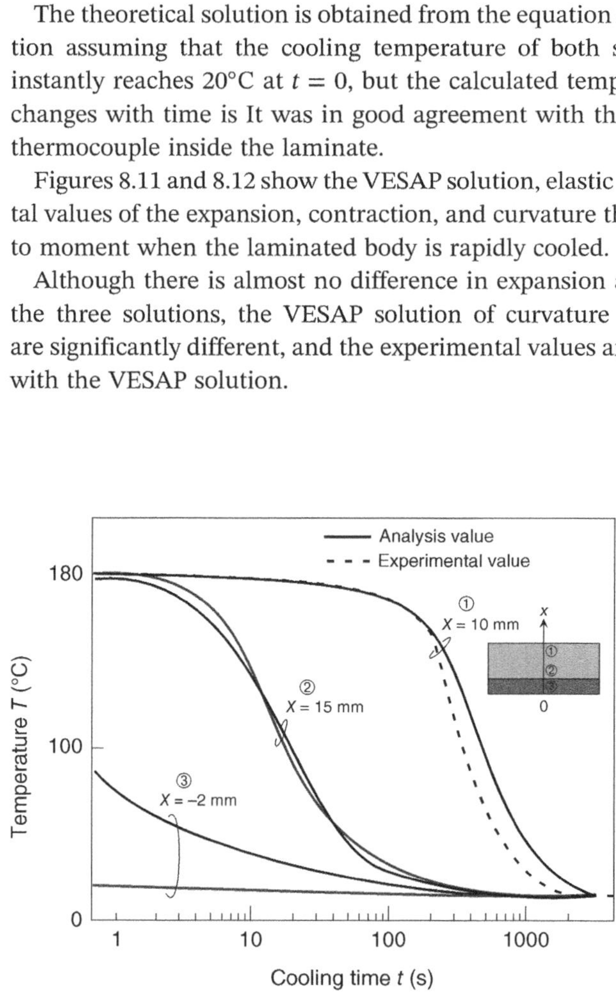

Figure 8.10 Temporal changes on the surface and inside.

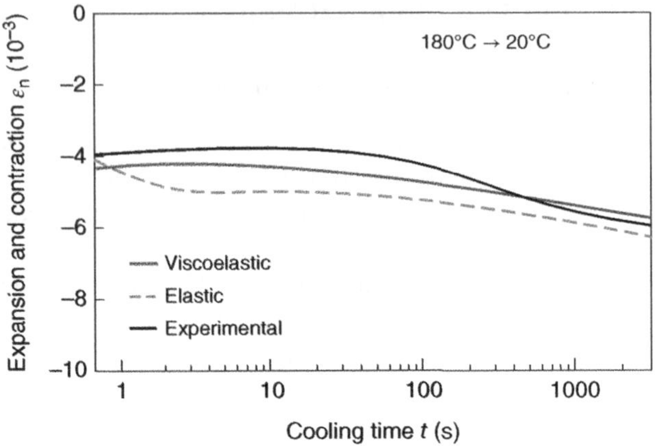

Figure 8.11 Expansion and contraction of analysis value.

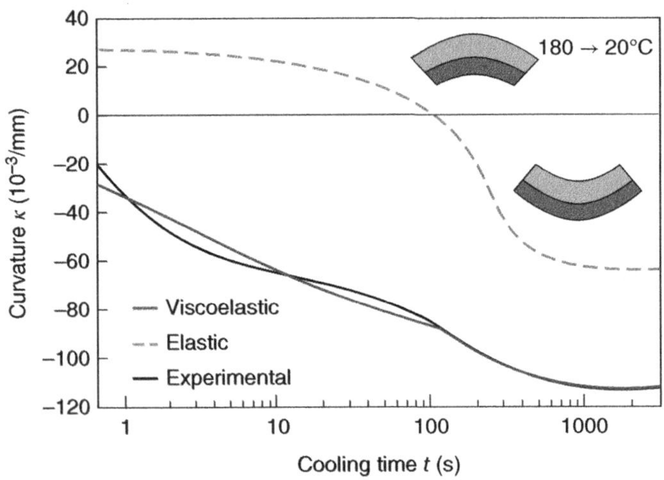

Figure 8.12 Theoretical and experimental values of curvature during quenching.

From these facts, it was clarified that the warp deformation in the quenching process of the laminated body is caused by the viscoelastic behavior of the constituent materials.

Although the slow cooling was omitted, the experimental values did not match the elastic solution, but the VESAP solution showed a good match, as in the case of quenching. From this, the validity of VESAP analysis can be confirmed.

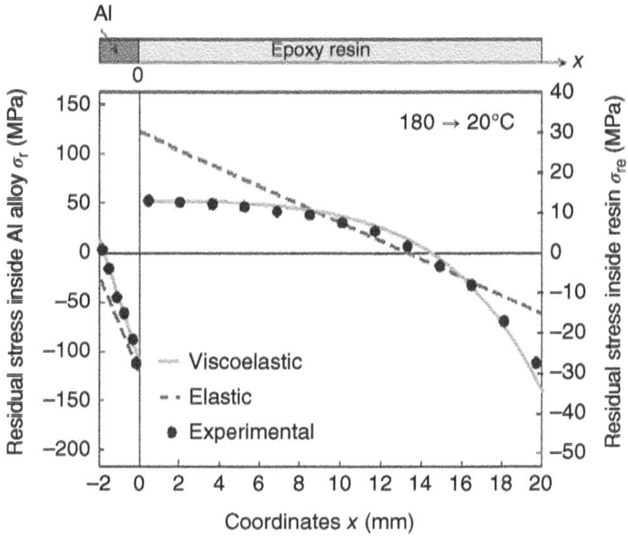

Figure 8.13 Theoretical and experimental values of residual stress.

Figure 8.13 shows the residual stress distribution of the laminate at room temperature after quenching in comparison with the results of VESAP analysis (hereinafter, viscoelastic solution), the results of elastic analysis by FEM (hereinafter, elastic solution), and experimental values.

As can be seen from Fig. 8.13, the viscoelastic solution is significantly different from the elastic solution, but it is in excellent agreement with the experimental values.

In addition, a large residual stress difference (corresponding to shear stress) occurs at the interface between the epoxy resin and the Al alloy, and the residual stress is the tensile stress near the junction interface, but it curves toward the surface to the residual stress of compression. It is characterized by its distribution.

8.5.2 In the Case of a Laminate of Encapsulating Resin/42 Alloy

The results of VESAP analysis (VESAP solution) of the amount of warpage deformation that occurs when a laminate consisting of LSI encapsulating resin and 42 alloy is cooled from a high temperature are compared with the elastic solution and experimental values, and their representative examples are shown in Figs. 8.14 and 8.15.

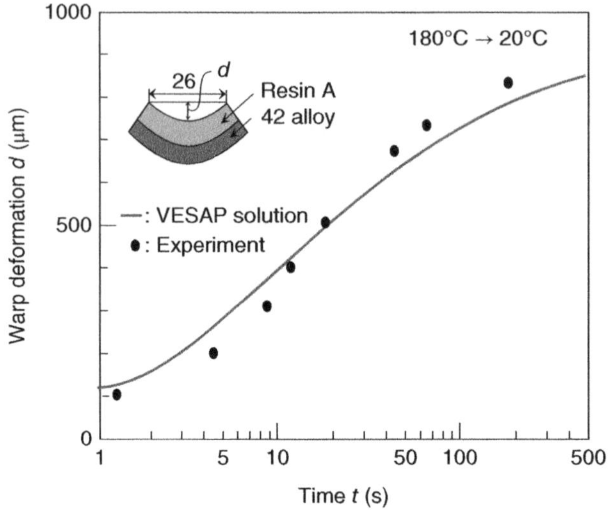

Figure 8.14 Warp deformation behavior during the cooling process (Resin A, thickness 1.25 mm).

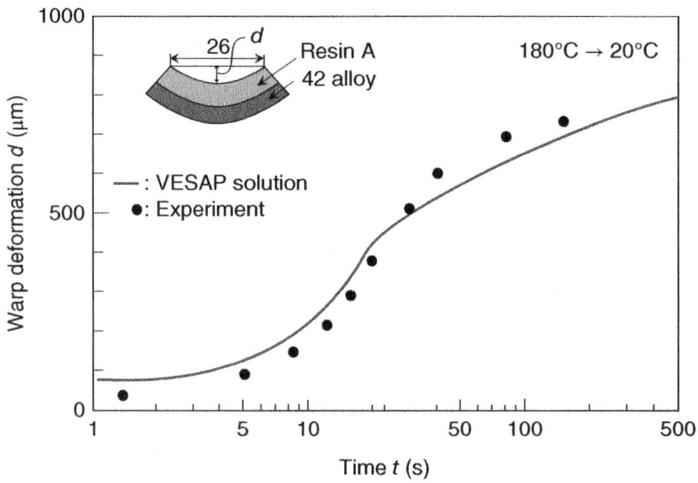

Figure 8.15 Warp deformation behavior during the cooling process (Resin B, thickness 1.25 mm).

Figure 8.14 shows the case of resin A with a high glass transition temperature, and Fig. 8.15 shows the case of resin B with a low glass transition temperature, and the thickness of the resin is 1.25 mm.

As can be seen from Figs. 8.14 and 8.15, the VESAP analysis on the warp deformation behavior of the laminate is in good agreement with the experimental values, but it is significantly different from the elastic solution. That is, the amount of warp deformation of the laminated body is almost zero at the time of molding at 175°C immediately after cooling, but warp deformation occurs due to the difference in the amount of heat shrinkage between the resin and 42 alloy with the passage of cooling time thereafter.

The amount of warpage deformation increases sharply up to the glass transition temperature T_g of the resin, but when the temperature is T_g or less, the tendency of increase in warpage deformation slows down, and the temperature of the laminate reaches 20°C after a sufficient cooling time. It converges to a constant value in the state where it becomes.

It is considered that this is because the coefficient of linear expansion of the resin at a temperature above T_g is about four times larger than the coefficient of linear expansion at a temperature below T_g.

8.6 Conclusion

As can be seen from the above, it was clarified that the amount of warp deformation and residual stress of the laminated body after cooling are caused by the viscoelastic behavior of the epoxy resin.

Furthermore, the results of VESAP analysis considering the viscoelasticity of the epoxy resin are significantly different from the results of the elastic analysis in the warp deformation behavior and residual stress distribution of the laminate, but they are in excellent agreement with the experimental values.

For the problem of dealing with body resin, we considered viscoelastic properties such as VESAP analysis instead of elastic analysis, but it is clear that analysis is indispensable.

Further, as described in the case of the laminate of epoxy resin and Al alloy, the cooling rate affects the deformation and residual stress of the laminate. Therefore, when designing the mounting of electronic equipment using an actual semiconductor package, it is necessary to pay attention not only to the viscoelastic properties of the resin but also to the thermal process during manufacturing.

9

Application to CSP-μBGA

9.1 Introduction

Up to this point, I have mainly discussed various issues related to semiconductor packages (Chapter 7). Next, using the viscoelastic analysis software (VESAP developed by the author) to solve these problems, I verified the validity and effectiveness of the most basic two-layer laminate model (Chapter 8).

In this chapter, as a specific example, a small and thin BGA-type semiconductor package, such as CSP-μBGA (Chip Scale Package-μBall Grid Array), which is widely used as a semiconductor package, is examined from the viewpoint of its constituent materials, structure, and thermal process.

9.2 Structure and Modeling of CSP-μBGA

The cross-sectional structure of the CSP-μBGA is shown in Fig. 9.1. This CSP-μBGA has a structure in which a soft elastomer is interposed between the LSI chip and the TAB tape to buffer the thermal stress. In addition, the ends of the TAB tape and Cu lead are sealed using a filler-filled epoxy resin.

Therefore, Visco-Elastic Stress Analysis Program (VESAP) analysis was performed to predict the thermal stress and warpage deformation behavior of this CSP-μBGA.

In the analysis, this was modeled into the multilayered laminate shown in Fig. 9.2. Here, Cu leads and encapsulants were excluded from the analysis because they do not become strength members.

In this laminate, we paid particular attention to the elastomer structure, which is expected to greatly affect the thermal stress and warpage deformation behavior of CSP-μBGA. Then, as the elastomer structure, three types of (a) structure A, (b) structure B and (c) structure C shown in Fig. 9.3 were taken up.

Polymeric Materials for Electronic Packaging, First Edition. Shozo Nakamura.

Published 2023 by John Wiley & Sons, Inc.

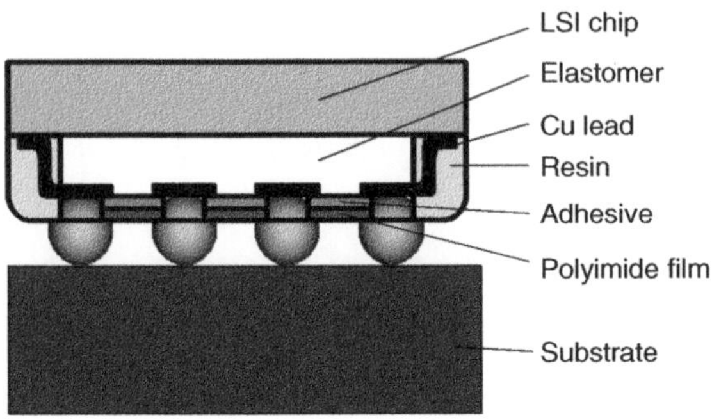

Figure 9.1 Cross section of CSP-μBGA.

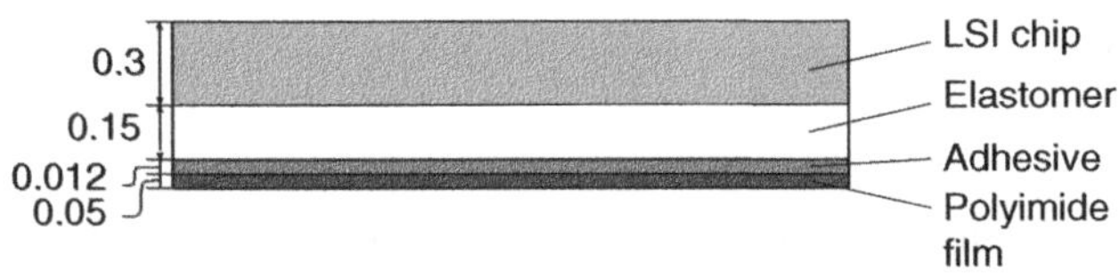

Figure 9.2 Modeled structure of CSP-μBGA.

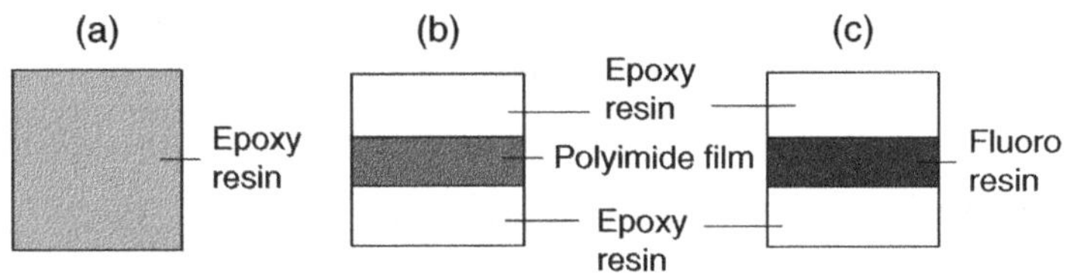

Figure 9.3 Structure of three kinds of elastomer. (a) Structure A, (b) Structure B, and (c) Structure C.

9.3 Material Property Values Used for Analysis

Table 9.1 shows the thermal and mechanical properties of the various materials that make up the laminate shown in Figs. 9.2 and 9.3.

Here, the epoxy resin, the adhesive, and the epoxy acrylate resin are organic materials, and their thermal and mechanical properties are greatly affected by temperature and time, and are materials that exhibit so-called remarkable viscoelastic behavior.

Therefore, for epoxy resins, adhesives, and epoxy acrylate resins, the dynamic viscoelastic properties were measured with the frequency as a variable parameter over a wide temperature range from low temperature to high temperature.

Table 9.1 Physical properties of GSP-μBGA constituent materials.

Properties / Materials	Thermal conductivity λ [W/(m K)]	Specific heat C (kJ/(kg K))	Density ρ (kg/m^3)	Glass transition temperature T_g (°C)	Thermal expansion coefficient α [10^{-6}/K]		Modulus E (GPa)	
					$T \leqq T_g$	$T_g < T$	$T \leqq T_g$	$T_g < T$
LSI chip	147.7	0.800	2328	—	3.6		192.0	
Epoxy resin	0.186	1.42	1220	89	88.0	217.0	3.4	0.03
Epoxy acrylate resin	0.180	1.38	1200	35	91.0	225.0	3.0	0.01
Fluororesin	0.162	1.09	1400	179	86.0	210.0	0.85	0.29
Adhesive	0.180	1.39	1220	90	70.0	161.0	3.3	0.03
Polyimide	0.162	1.09	1431	420	25.0	47.2	4.3	1.75

This measurement result was plotted on the converted logarithmic time axis under the reference temperature T_0, and a master curve of the relaxation modulus was created.

Here, as an example, the master curve of the relaxation elastic modulus of the epoxy resin is shown in Fig. 9.4.

As can be seen from Fig. 9.4, the relaxation elastic modulus of the epoxy resin can be obtained as a single smooth master curve from a very short time to a very long time.

In addition, as shown in Fig. 9.5, the time–temperature transfer factor when creating this master curve could be approximated by two Arrhenius equations with different activation energies ΔH in the high-temperature part and the low-temperature part.

Although omitted here, the same master curve as the epoxy resin was obtained for the adhesive and the epoxy acrylate resin.

From this, it was found that the epoxy resins and adhesives taken up here are materials that show remarkable viscoelastic behavior and that the time–temperature conversion rule holds. These values were used for VESAP analysis.

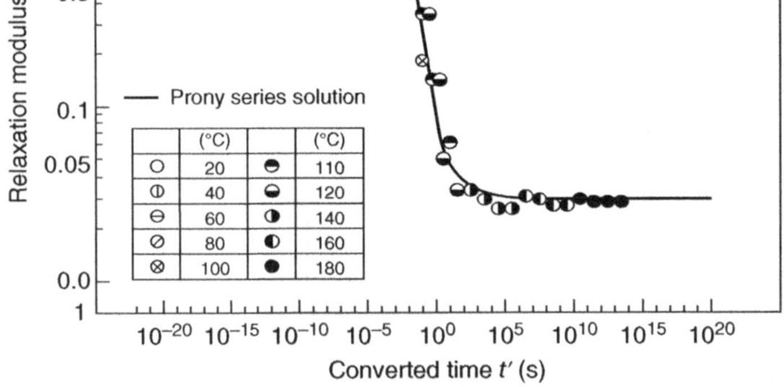

Figure 9.4 Master curve of epoxy resin.

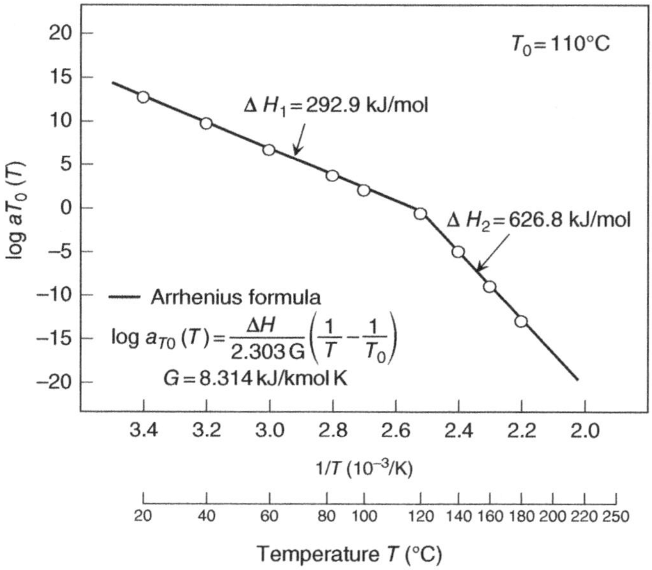

Figure 9.5 Time–temperature shift factor of epoxy resin.

9.4 Material and Structure Optimization Design by VESAP Analysis

A series of analysis results on the interfacial thermal stress difference analyzed by viscoelasticity for the laminate modeled on CSP-μBGA are shown in Fig. 9.6.

Figure 9.6 focuses on the interface between the LSI chip and the elastomer, which is expected to cause the greatest damage to the CSP-μBGA due to the heat load during the temperature cycle test, and analyzes the interface thermal stress difference for the cooling time from 150°C.

Here, the interfacial thermal stress difference defines the thermal stress difference between the thermal stress generated on the LSI chip side and the thermal stress generated on the elastomer side at the connection interface between the LSI chip and the elastomer, and this thermal stress difference is the delamination.

From Fig. 9.6, the difference in interfacial thermal stress increases with the passage of cooling time for structure A, structure B, and structure C of the elastomer, and is constant when the CSP-μBGA reaches −55°C after 50 seconds of cooling time. Converges to the value.

Here, looking at the types of elastomer structures, the newly proposed structure C made of epoxy resin/fluororesin/epoxy resin is compared with the conventional structure A and structure B made of epoxy resin single layer and

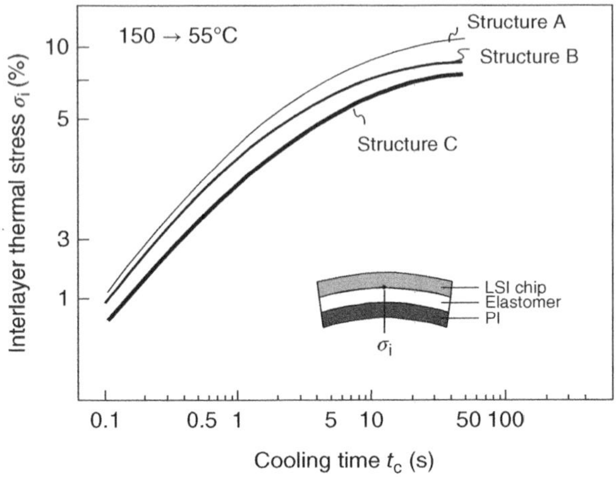

Figure 9.6 Change of thermal stress during the cooling process.

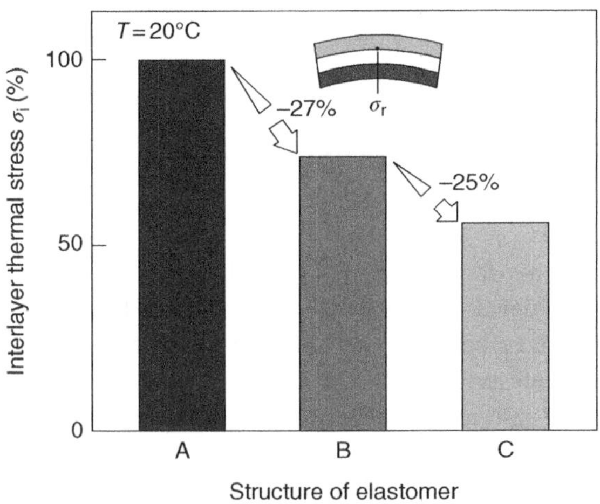

Figure 9.7 Comparison of thermal stress by structure.

epoxy resin/PI/epoxy resin. It can be seen that the difference in surface thermal stress is smaller in all cooling time zones.

On the other hand, Fig. 9.7 shows the results of comparing the interfacial thermal stress difference between the LSI chip and the elastomer at room temperature of 20°C, focusing on the differences between the structures A, B, and C of the elastomer. From Fig. 9.7, it can be seen that the interfacial thermal stress difference is significantly reduced by 25–45% in the newly proposed structure C compared to the conventional structures A and B at room temperature of 20°C.

This is due to the material properties of the rigid LSI chip showing elastic behavior and the soft elastomer showing viscoelastic behavior, which largely depends on the heat shrinkage force based on the difference between the linear expansion coefficient and elastic modulus of both materials.

That is, in the structures A and B, the epoxy resin, which is soft and has a large coefficient of linear expansion, has a large proportion in the entire elastomer, whereas the newly proposed structure C has a small ratio, and the bending rigidity and heat of the entire laminate are large. It is considered that the contraction force is a result of complicated and mutual influence.

On the other hand, the residual stress distribution generated inside this laminate is shown in Fig. 9.8 for the structure A, B, and C of the elastomer.

Figure 9.8 shows the residual stress distribution that occurs at −55°C when air-cooled from 150°C to −55°C in the temperature cycle test conducted during the reliability test.

Characteristically, these results show that tensile residual stress is generated on the surface of the LSI chip, and compressive residual stress is generated at the interface with the epoxy resin inside, whereas elastomers, adhesives, and polyimides are generated. Residual tensile stress is generated inside (PI).

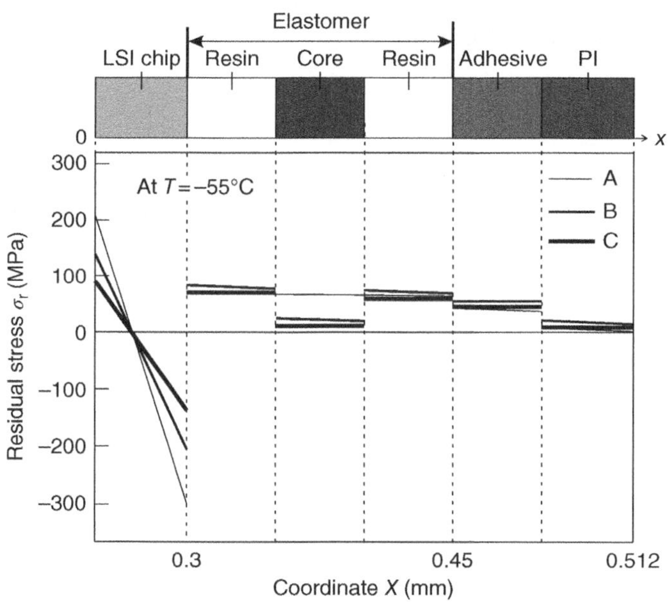

Figure 9.8 Residual stress distribution of CSP-μBGA.

In this way, CSP-μBGA, which is composed of a wide variety of materials such as LSI chips, elastomers, adhesives, and PIs, has a complicated residual stress distribution inside, which influence each other and in a complicated manner.

By interacting with each other, it leads to defective morphology such as disconnection of Cu leads and delamination of the member interface.

In particular, a large shear stress is generated at the interface between the LSI chip and the epoxy resin of the elastomer, and cracks occur on the surface of the LSI chip from this portion, or interface peeling occurs when the adhesive strength between the two is low.

From this point of view, the structure C is less likely to break the LSI chip or peel off the interface than the conventional structures A and B, and is considered to be superior in terms of connection reliability.

Figure 9.9 shows the VESAP analysis results of the warp deformation behavior that occurs when the above-mentioned laminate is cooled from 150°C to −55°C in the temperature cycle test.

From Fig. 9.9, it can be seen that the amount of warpage deformation of the laminated body increases remarkably with the passage of cooling time.

Figure 9.10 shows the residual curvature (means residual warpage deformation) by elastomer structure.

Focusing on the difference in elastomer structure, it can be said that the newly proposed structure C has a significantly smaller amount of warp deformation than the conventional structure A and structure B, and is superior in terms of the reliability of the connection strength of CSP-μBGA.

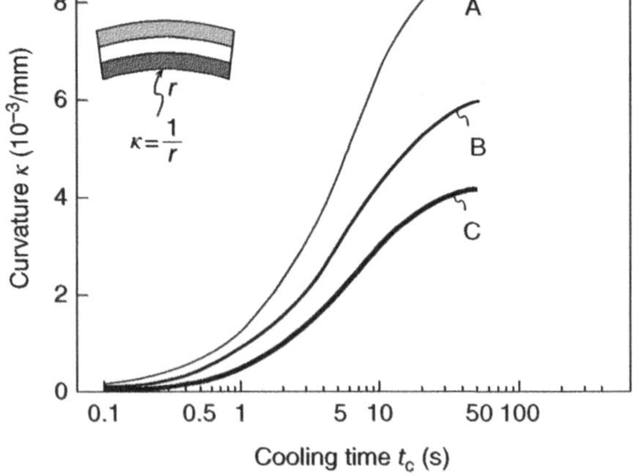

Figure 9.9 Warp deformation behavior during the cooling process of CSP-μBGA.

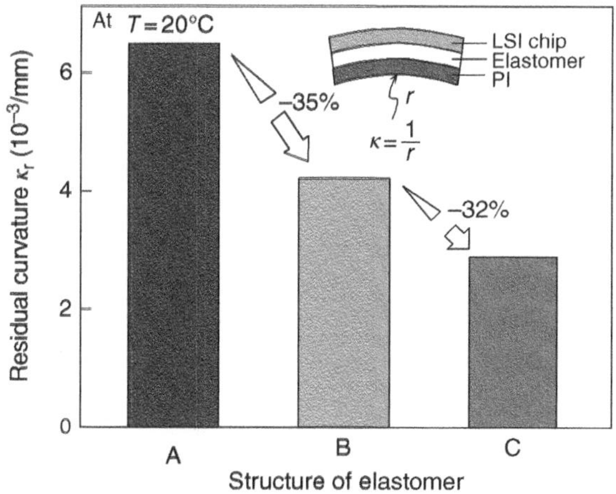

Figure 9.10 Comparison of residual curvature by structure.

From this, it can be seen that the structure C is superior to the structure A and the structure B in terms of warpage deformation as well as the above-mentioned residual stress.

In this way, the complex residual stress distribution and warp deformation behavior inside the material are due to the flexural rigidity of structures with different physical characteristics of various materials and the thermal expansion/contraction force due to the difference in the linear expansion coefficient and elastic modulus of the material. Is the result of complex influences on each other.

Therefore, it prevents interfacial peeling and cracking between the constituent materials of CSP-μBGA that are subject to various heat loads and causes electricity. In order to improve the connection strength as a slave device and the reliability of warpage deformation, optimized design of constituent materials and structures is indispensable.

In the end, although omitted here, the reliability life evaluation of CSP-μBGA alone was performed. As a result, all items passed and were shipped to the market as a product.

From the results described above, it can be seen that the VESAP analysis developed by the author enables the optimized design of materials and structures of electronic components such as LSI packages.

10

Thermal Stress and Warpage Behavior During Cooling Process of Three-Layer Laminate

10.1 Introduction

In Chapter 8, I explained the case where the two-layer laminate was slowly cooled and rapidly cooled. Viscoelasticity analyses were performed using VESAP, and these results were obtained. I stated that the validity and effectiveness of the VESAP analysis could be verified by comparing it with the experimental values.

In this chapter, the basic formula of viscoelasticity analysis by VESAP derived was used and applied to a general-purpose LSI package, which is a practical problem.

Specifically, we conducted a VESAP analysis on a three-layered laminate that is close to a general-purpose semiconductor package shape and has a basic shape and examined practical problems through a material parameter survey.

10.2 Structure of LSI Package

As mentioned in Chapter 1, LSI packages are roughly classified into "lead insertion type" and "surface mount type" according to the mounting method, and there are packages with shapes and structures corresponding to the mounting method.

For example, the lead insertion type has DIP (Dual Inline Package) and PGA (Pin Grid Array), and the surface mount type has QFP (Quad Flat Package) and BGA (Ball Grid Array), etc., which are typical.

The LSI package has various functions such as supplying power to the LSI chip and transmitting the signal emitted by the LSI chip to other devices, and sealing it with resin to protect it from the external environment. The LSI chip does not function as it is, but it connects the LSI chip to the outside to enable power supply and electrical signal input/output. LSI packages play a role in maximizing the capabilities of LSI chips.

Polymeric Materials for Electronic Packaging, First Edition. Shozo Nakamura.

Published 2023 by John Wiley & Sons, Inc.

A typical LSI package is DIP, which is a structure in which an LSI chip is covered with sealing resin. That is, since the top and bottom of the LSI chip can be regarded as a resin structure, it can be treated as a three-layer structure model of resin/Tab LSI chip/resin.

As mentioned earlier, silicon (Si) is used for the LSI chip, which is a material having high rigidity and an extremely small coefficient of linear expansion. Since a very fine circuit is formed on the LSI chip, it is sealed with resin to protect it from the external environment such as external force, dust, and humidity.

This encapsulating resin or elastomer has a resin configuration that is as close to the linear expansion coefficient of the LSI chip as possible in order to eliminate defects such as peeling of the adhesive interface and resin cracks due to thermal expansion and contraction when heat is applied by soldering.

In addition, at the interface between dissimilar materials, problems such as peeling and resin cracks occur. At these interfaces, so-called stress and strain singular fields are created, and stress and strain analysis in the singular field region becomes a fracture mechanics problem, which is more complicated.

In recent years, the dimensions of LSI chips have increased with the increase in the capacity of semiconductor elements in semiconductor packages, but the outer shape of semiconductor packages tends to be smaller and thinner, resulting in a weak structure in terms of moisture resistance and thermal deformation. Ensuring reliability has become an increasingly important technical issue.

10.3 Three-Layer Viscoelastic Laminate Model

As mentioned earlier, the interface peeling between the tab lead or LSI chip, the resin was one of the major factors that hindered the reliability of the semiconductor package.

Therefore, in order to evaluate the thermal stress generated at the interface between the tab lead or LSI chip and the resin.

Assuming such a structure, a three-layer laminate model that assumes the interface between resin and tab lead or LSI chips is generally conceived. Therefore, assuming a three-layer laminate, I considered and examined the laminate model shown in Fig. 10.1.

Figure 10.1 shows the tab lead/resin interface, LSI chip/resin interface, etc., as the most basic laminate model assuming the above interface.

A material parameter survey was conducted focusing on the warp deformation behavior due to thermal stress and heat load, focusing on the three-layered laminate.

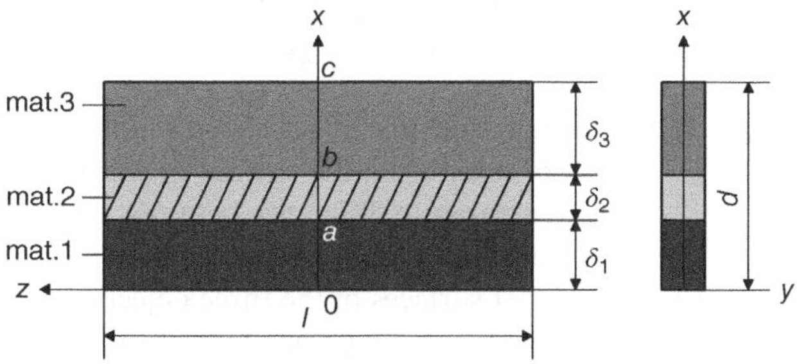

Figure 10.1 Modeled three-layer laminate.

10.4 Elucidation of Warpage Deformation Behavior by VESAP Analysis

10.4.1 Calculation Formula of Temperature Distribution

Since semiconductor packages are made of materials with different properties, various heats are applied during the manufacturing process. Most of the causes of defects such as interface peeling of semiconductor packages, resin cracks, and warpage deformation are due to heat load.

Therefore, it is necessary to investigate what the temperature inside the semiconductor package is due to the heat loaded during manufacturing. If the temperature distribution is known, it is possible to predict where in the package the defect will occur.

It is practically impossible to actually measure the temperature distribution inside this package, and it is generally obtained by numerical analysis. Here, the temperature distribution is calculated by thermal analysis for the laminate modeled on the package.

Now, when the modeled laminate is cooled from the high temperature at the time of manufacture to room temperature, the temperature inside the laminate decreases with time, and the temperature is expressed as a function of location x and time t.

Therefore, the basic equation for obtaining the temperature distribution $U(x, t)$ inside the laminate, which changes from moment to moment, is derived based on the following assumptions.

Here, assuming a three-layer laminate shown in Fig. 10.2, the initial temperature of the laminate is assumed to be Θ_1, and the temperature of both sides is assumed to be Θ_2.

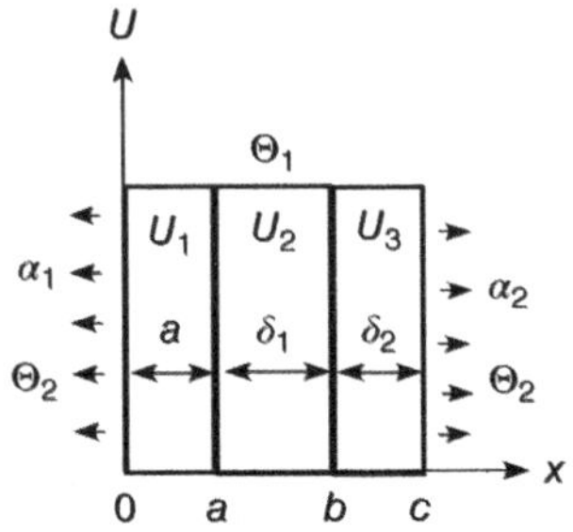

Figure 10.2 Schematic diagram of temperature distribution.

Let α_1 be the heat transfer coefficient on the $x = 0$ plane, and α_2 be the $x = c$ plane.

i) On both surfaces in contact with the outside air, heat is released from the surface in proportion to the temperature difference between the outside air and the surface by heat transfer.
ii) The contact surfaces of the three objects are in perfect contact, and the temperature and amount of heat flowing between them are the same.
iii) Material property values of an object (e.g. specific heat, thermal conductivity, and specific weight) are constant regardless of temperature.

Assuming that the temperature of the object is $T(x, t)$ and the surface of object 1 is the coordinate origin, the differential equation and conditional equation of one-dimensional unsteady heat conduction can be expressed by the following equations.

Note that I represents the type of material constituting the laminate.

$$\frac{\partial}{\partial t}U_i(x,t) = \xi_i \frac{\partial^2}{\partial x^2}U_i(x,t) \quad i = 1, 2, 3 \tag{10.1}$$

The following assumptions generally hold as the initial and boundary conditions. That is, before applying heat to the semiconductor package, that is, when the time t is 0, the temperature is the same for all three types of materials, and the temperature is the initial temperature T_b, and at the interface where the three materials are in contact. The following equation is obtained from various conditions such as heat transfer.

At,

$t = 0$,

$$U_1(x,0) = U_2(x,0) = U_3(x,0) = \Theta_1 \tag{10.2}$$

$x = 0$,

$$\frac{\partial}{\partial x}U_1(0,t) - h_1\{U_1(0,t) - \Theta_2\} = 0 \tag{10.3}$$

$x = a$,

$$U_1(a,t) = U_2(a,t) \tag{10.4}$$

$$\lambda_1 \frac{\partial}{\partial x} U_1(a,t) = \lambda_2 \frac{\partial}{\partial x} U_2(a,t) \tag{10.5}$$

$x = b$,

$$U_2(b,t) = U_3(b,t) \tag{10.6}$$

$$\lambda_2 \frac{\partial}{\partial x} U_2(b,t) = \lambda_3 \frac{\partial}{\partial x} U_3(b,t) \tag{10.7}$$

$x = c$,

$$\frac{\partial}{\partial x} U_3(c,t) + h_2\{U_3(c,t) - \Theta_2\} = 0 \tag{10.8}$$

Here, $h_1 = \frac{\alpha_1}{\lambda_1}$, $H_2 = \frac{\alpha_2}{\lambda_2}$.

Therefore, when Eq. (10.1) is solved so as to satisfy the conditions of the given equations, the temperature distributions $U_1(x,t)$, $U_2(x,t)$, and $U_3(x,t)$ of the desired object 1, 2, 3, are obtained. They are finally expressed by the following equation:

$$U_1 = \Theta_2 + 2(\Theta_1 - \Theta_2)\sum_{n=1}^{\infty}\frac{E_x}{\mu_n W(\mu_n)}\left[H_a\left\{G(\mu_n)\cos\frac{a-x}{a}\mu_n + H(\mu_n)\sin\frac{a-x}{a}\mu_n\right\}\right.$$
$$\left. + \frac{H_c}{p}K_2K_3\left\{\mu_n\cos\frac{x}{a}\mu_n + H_a\sin\frac{x}{a}\mu_n\right\}\right] \tag{10.9}$$

$$U_2 = \Theta_2 + 2(\Theta_1 - \Theta_2)\sum_{n=1}^{\infty}\frac{E_x}{\mu_n W(\mu_n)}\left[H_aK_1\left\{E(\mu_n)\cos\frac{b-x}{\delta_2}m\mu_n + F(\mu_n)\sin\frac{b-x}{\delta_2}m\mu_n\right\}\right.$$
$$\left. + \frac{H_c}{p}K_3\left\{A(\mu_n)\cos\frac{x-a}{\delta_2}m\mu_n + B(\mu_n)\sin\frac{x-a}{\delta_2}m\mu_n\right\}\right] \tag{10.10}$$

$$U_3 = \Theta_2 + 2(\Theta_1 - \Theta_2)\sum_{n=1}^{\infty}\frac{E_x}{\mu_n W(\mu_n)}\left[H_aK_1K_2\left\{\mu_n\cos\frac{c-x}{\delta_3}p\mu_n + \frac{H_c}{p}\sin\frac{c-x}{\delta_3}p\mu_n\right\}\right.$$
$$\left. + \frac{H_c}{p}\left\{C(\mu_n)\cos\frac{x-b}{\delta_3}p\mu_n + D(\mu_n)\sin\frac{x-b}{\delta_3}p\mu_n\right\}\right] \tag{10.11}$$

Here,

$$\delta_2 = b - a, \quad \delta_3 = c - b, \quad H_a = h_1 a, \quad H_c = h_2\delta_3$$
$$K_i = \sqrt{\lambda_i c_i \rho_i}\ (i = 1, 2, 3)$$
$$m = \frac{\delta_2}{k_2}\frac{k_1}{a}, \quad p = \frac{\delta_3}{k_3}\frac{k_1}{a}, E_x = \exp\left\{-\left(\frac{k_1\mu_n}{a}\right)^2 t\right\}$$

$$W(\mu_n) = E(\mu_n)\bar{N}(\mu_n) + \bar{F}(\mu_n)M(\mu_n) - \bar{E}(\mu_n)N(\mu_n) - F(\mu_n)\bar{M}(\mu_n)$$

$$E(\mu_n) = K_2\left(\mu_n \cos p\mu_n + \frac{H_c}{p}\sin p\mu_n\right)$$

$$F(\mu_n) = K_3\left(\frac{H_c}{p}\cos p\mu_n - \mu_n \sin p\mu_n\right)$$

$$M(\mu_n) = A(\mu_n)\cos m\mu_n + B(\mu_n)\sin m\mu_n$$

$$N(\mu_n) = B(\mu_n)\cos m\mu_n - A(\mu_n)\sin m\mu_n$$

$$A(\mu_n) = K_2(\mu_n \cos \mu_n + H_a \sin \mu_n)$$

$$B(\mu_n) = K_1(H_a \cos \mu_n - \mu_n \sin \mu_n)$$

$$\bar{E}(\mu_n) = K_2\{(1 + H_c)\cos p\mu_n - p\mu_n \sin p\mu_n\}$$

$$\bar{F}(\mu_n) = K_3\{p\mu_n \cos p\mu_n + (1 + H_c)\sin p\mu_n\}$$

$$\bar{M}(\mu_n) = \{\bar{A}(\mu_n) + mB(\mu_n)\}\cos m\mu_n - \{mA(\mu_n) + \bar{B}(\mu_n)\}\sin m\mu_n$$

$$\bar{N}(\mu_n) = \{mA(\mu_n) + \bar{B}(\mu_n)\}\cos m\mu_n + \{\bar{A}(\mu_n) + mB(\mu_n)\}\sin m\mu_n$$

$$\bar{A}(\mu_n) = K_2\{(1 + H_a)\cos \mu_n - \mu_n \sin \mu_n\}$$

$$\bar{B}(\mu_n) = K_1\{\mu_n \cos \mu_n + (1 + H_a)\sin \mu_n\}$$

$$C(\mu_n) = K_3M(\mu_n), D(\mu_n) = K_2N(\mu_n)$$

$$G(\mu_n) = K_1\{E(\mu_n)\cos m\mu_n + F(\mu_n)\sin m\mu_n\}$$

$$H(\mu_n) = K_2\{F(\mu_n)\cos m\mu_n - E(\mu_n)\sin m\mu_n\}$$

Now,

μ_n is the nth root of $E(x)N(x) + F(x)M(x) = 0\ (n = 1, 2, 3\ldots)$

$$E(x) = K_2\left\{x \cos px + \frac{H_c}{p}\sin px\right\}$$

$$F(x) = K_3\left\{\frac{H_c}{p}\cos px - x \sin px\right\}$$

$$M(x) = A(x)\cos mx + B(x)\sin mx$$

$$N(x) = B(x)\cos mx - A(x)\sin mx$$

$$A(x) = K_2(x \cos x + H_a \sin x)$$

$$B(x) = K_1(H_a \cos x - x \sin x)$$

where, λ: thermal conductivity, ρ: density, c: specific heat, α: heat transfer coefficient

$$h = \alpha/\lambda,\ \ H_a = ha,\ \ H_b = hb.$$

10.4.2 Material Properties and Analysis Conditions

In the VESAP analysis, the physical property values in Table 10.1 and the master curves of the relaxation modulus of Resin A and Resin C are approximated by Prony series in Fig. 10.3, respectively, and the time–temperature shift factor shown in Fig. 10.4, respectively.

Table 10.1 Materials property of a laminated structure.

Property \ Material	Resin A	Resin B	Resin C	Cu	Fe–Ni alloy: 42 alloy	Fe–Ni alloy: 50 alloy
λ (kcal/m h °C)	0.677	0.652	0.742	133.2	13.7	13.3
C (kcal/kg °C)	0.220	0.211	0.196	0.09	0.11	0.11
ρ (kg/m^3)	1823	1832	1940	8600	8120	8300
T_g (°C)	135	132	158	—	—	—
α_1 (10^{-5}/°C)	1.70	1.72	1.00	1.70	0.58	1.00
α_2 (10^{-5}/°C)	5.46	6.24	3.99	—	—	—
E (GPa)	1.3–1.12	2.1–21.0	7.7–22.7	115	150	168

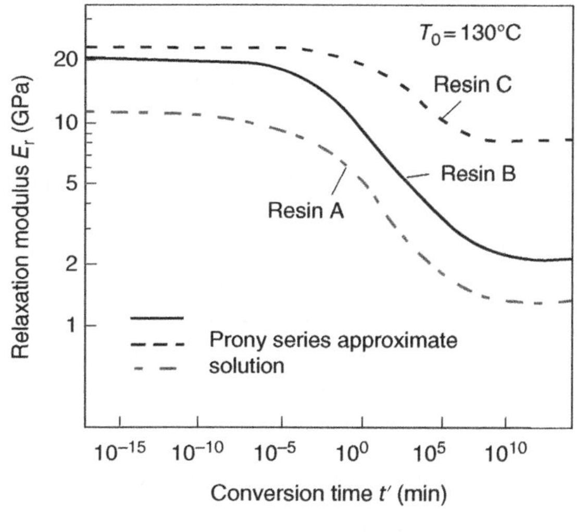

Figure 10.3 Relaxation modulus of resins for various LSI packages.

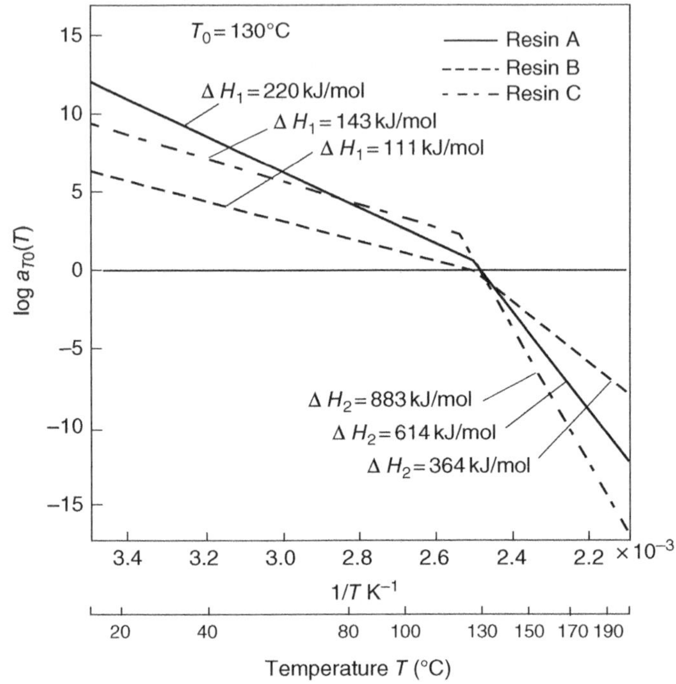

Figure 10.4 Time–temperature transfer factor of various resins.

The time–temperature shift factor was approximated by two Arrhenius equations with different activation energies ΔH in the high-temperature part and the low-temperature part, and these values were used.

As shown in Table 10.1, two types of metal materials for lead frames, 50 alloy (50 alloy) and 42 alloy (42 alloy), which are Sn/Ni-plated Cu and Fe–Ni alloys, were used.

On the other hand, the temperature conditions for VESAP analysis were the case where the temperature was maintained at a high temperature holding temperature of 180°C, which is almost equal to the molding temperature, and then cooled to the room temperature of 20°C. The cooling rate at this time was assumed to be the case where the semiconductor package with the lead frame was air-cooled as described above, and a heat transfer coefficient of 40 kcal/m^2 h °C was adopted. The VESAP analysis was performed by the same method as described earlier.

10.4.3 Analysis Results and Discussion

In the analysis of thermal stress and warpage deformation generated in the three-layer laminate, typical examples are (1) resin A/Cu lead/resin A, (2) resin C/50 alloy/resin C, and (3) resin C/Cu lead/resin C.

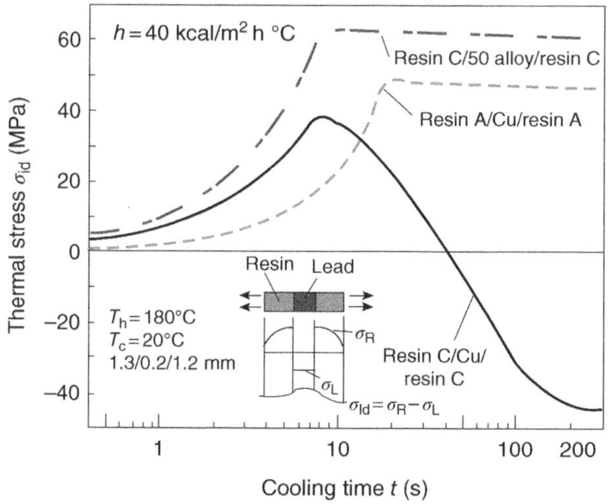

Figure 10.5 Thermal stress changes of laminate.

As an analysis results, Fig. 10.5 shows the interfacial thermal stress at the interface that changes from moment to moment during cooling, and Fig. 10.6 shows the curvature change in the three laminates.

The interfacial thermal stress represents the difference between the thermal stress on the resin side and the thermal stress on the lead side near the interface between the resin and the lead. This means that peeling will occur at the interface between two materials.

As can be seen from Fig. 10.5, in the combination of (1) resin A/Cu lead/resin A and (2) resin C/50 alloy/resin C, the interfacial thermal stress increases monotonically with cooling and then becomes a constant value for 200 seconds.

Figure 10.6 Curvature changes of laminate.

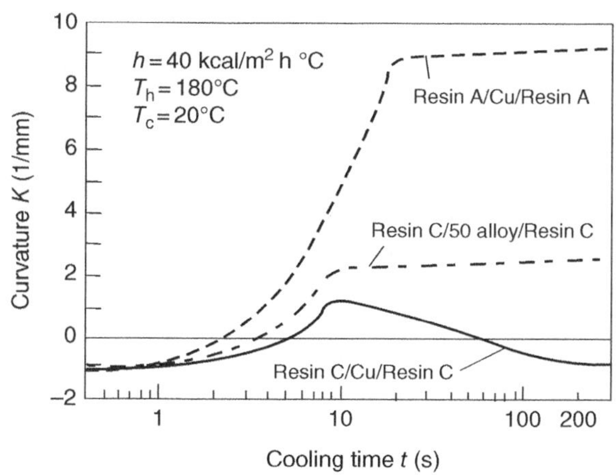

It can be seen that the room temperature reaches 20°C. That is, at room temperature, there is an interfacial thermal stress of about 50 MPa in the combination of (1) and about 60 MPa in the combination of (2) at the interface between the resin and the lead.

Here, the time at which the interfacial thermal stress becomes a constant value in the combination of the two corresponds to the time at which the resin temperature becomes exactly equal to the glass transition temperature.

That is, below the glass transition temperature of the resin, since the linear expansion coefficients of the lead and the resin are the same, the difference in thermal stress between the two does not occur and shows a constant value.

On the other hand, the combination of (3) resin C/Cu lead/resin C is different from the behavior of the interfacial thermal stress of the other two combinations (1) and (2).

That is, in the case of (3), the positive interfacial thermal stress is shown at the initial stage of cooling, but as the cooling progresses over time, it reverses to the negative interfacial thermal stress, the so-called double swing.

As can be seen from the change in curvature shown in Fig. 10.6, in the case of combinations (1) and (2), both are warped and deformed in a convex state, but in the case of (3).

It can be seen that the warp is convex upward at first, then the warp decreases and becomes flat, and then the warp deformation is convex downward and reaches room temperature.

Figure 10.7 shows the residual stress distribution generated inside the three-layer laminate in these three types of resin/lead/resin combinations. Here, the thermal stress of each laminated body at room temperature is called residual stress.

From Fig. 10.7, the combination of (1) resin A/Cu lead/resin A and (2) resin C/50 alloy/resin C shows the residual stress of tension inside the resin and the residual stress of compression inside the lead.

On the other hand, the combination of resin C/Cu lead/resin C, which has different linear expansion coefficients for both materials, shows a residual stress distribution opposite to that of these two combinations. In addition, it can be seen that the residual stress inside the laminate shows an almost uniform distribution in all the materials.

For reference, Fig. 10.8 shows the temperature change in the laminate. From this, the heat transfer coefficient was calculated.

As described earlier, in the case of a three-layer laminate, the temporal change behavior of the thermal stress generated at the interface includes the difference in the linear expansion coefficient between the lead and the resin and the difference in the thermal conductivity inside the material of the laminate.

It can be seen that it differs greatly depending on the type. For this reason, in the design and development of general LSI packages, there is an optimal combination

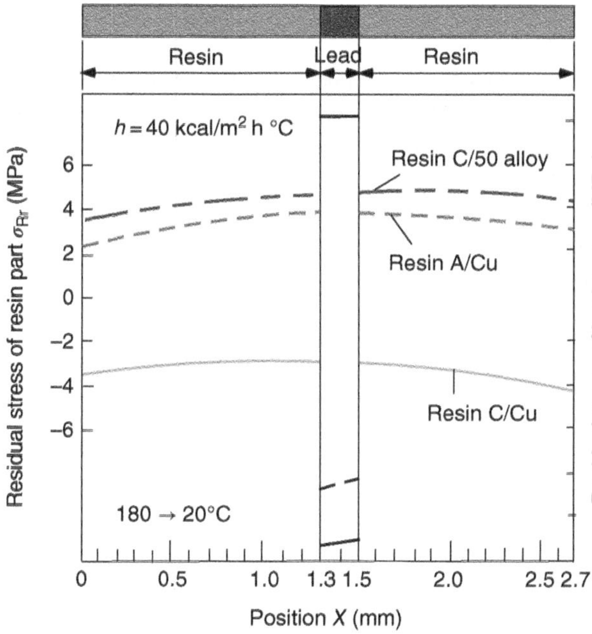

Figure 10.7 Residual stress distribution inside the laminate.

Figure 10.8 Temperature change of laminate.

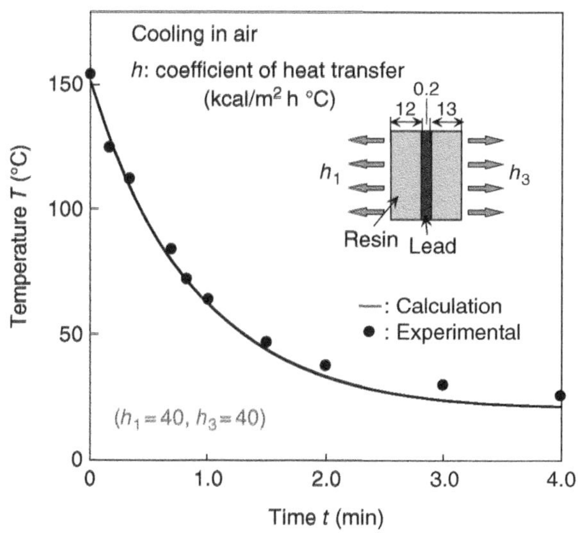

of constituent materials for reducing thermal stress and warpage deformation that cause interface peeling.

What was generally found from this basic analysis is that the selection of materials with no difference in the coefficient of linear expansion of the constituent materials and the glass transition temperature T_g of the resin greatly affect the amount of warpage deformation.

Therefore, it was shown that it is possible to optimize material design, thermal process design, and shape/dimension design by performing VESAP analysis considering the viscoelastic behavior of the resin in advance at the design stage.

11

Warp Deformation Behavior From Heating to Cooling

11.1 Introduction

Up to Chapter 10, I have verified that viscoelasticity analysis using Visco-Elastic Stress Analysis Program (VESAP) is valid and effective by comparing with experimental values for various laminates assuming parts for electronic devices such as semiconductor packages.

I have been arguing that VESAP analysis is useful for obtaining optimized design guidelines for materials, structures, and thermal processes of electronic devices, and is also a great weapon for shortening the development period of new products and technologies and reducing manufacturing costs.

Further, in recent years, in order for design engineers of companies to realize multifunctional and high-performance electronic devices and cost reduction, it is an issue to prevent warpage deformation of laminated boards having a large area.

Electronic devices such as multifunction mobile phones are becoming lighter, thinner, shorter, and smaller. Electronic components used in electronic equipment are generally composite structures composed of different materials such as IC chips, copper foils, printed circuit boards, and sealing resins. Therefore, when various thermal loads are applied in the manufacturing process, large thermal stress and strain are generated due to differences in material properties, which can cause defects such as interfacial peeling, material cracks, and warpage deformation. In order to prevent these defects, it is necessary to optimize the physical properties and structures of the materials that make up the electronic parts, as well as the manufacturing processes.

In recent years, there have been reports on the optimization of the manufacturing process, but no report on the heat load during cooling in the manufacturing process has been found.

Polymeric Materials for Electronic Packaging, First Edition. Shozo Nakamura.

Published 2023 by John Wiley & Sons, Inc.

However, it is considered to be an important research topic because warping deformation occurs when cooling from high temperatures in the manufacturing process. Therefore, in this study, the effects of cooling conditions on the warpage deformation behavior of viscoelastic laminates made of epoxy resin/aluminum that modeled electronic components were investigated both experimentally and theoretically.

11.2 Two-Layer Laminate With Epoxy Resin/FR-4 Substrate

Innumerable combinations are possible for the laminated body. The structure in which an insulating resin is coated on the large-scale integration (LSI) chip, the LSI chip is mounted on a glass/epoxy resin substrate via a resin such as underfill, and the elastomer for stress buffering is used.

The important thing in development is as follows:

i) To model the actual product to be developed into a simple and appropriate structure and shape.
ii) Being able to quantitatively and accurately predict values such as the amount of warpage deformation.
iii) It is important that anyone can easily analyze the analysis without the preparation and method being too complicated.

Nearly infinite mesh division, incorrect element selection, complicated input of material properties, and use only by those who require special skills require labor and speed without benefit. It hinders design and development.

From this point of view, this section takes up the measurement of a two-layered laminate that is a combination of viscoelastic materials that are considered to be the most complicated and are close to the real products.

As the material used, a general-purpose resin or the like that can be easily obtained was used. In the actual product manufactured by the manufacturer, it is only necessary to input the viscoelastic properties of the material used by the manufacturer at the time of VESAP analysis.

11.2.1 Experimental Method

Figure 11.1 shows the shape and dimensions of the laminate. The laminate is a two-layer laminate made of epoxy resin/aluminum, which is the simplest form of electronic components.

The epoxy resin is a bisphenol A type two-liquid mixed type that mixes the main agent and the curing agent at a ratio of 5:1.

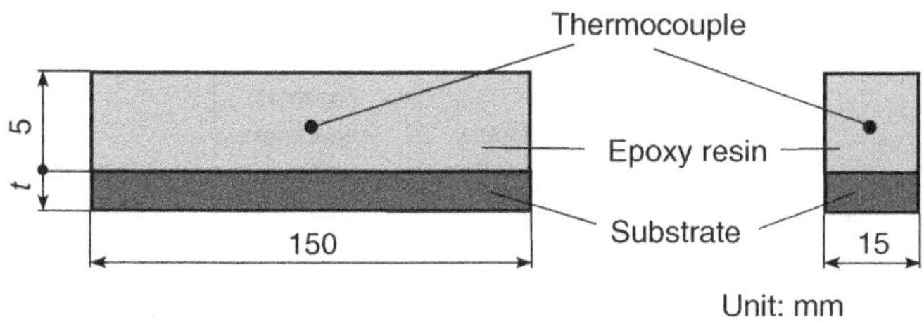

Figure 11.1 Laminate for epoxy resin and substrate.

On the other hand, AL1050P was used for aluminum. The laminate had a length of 150 mm, a width of 15 mm, an epoxy resin thickness of 5 mm, and an aluminum thickness of 1 mm. When bonding these materials together, an adhesive made of the same material as the epoxy resin was used, and the materials were left at room temperature of about 25°C for 24 hours for adhesive bonding. The same laminate as used in the experiment was used as a dummy, and thermocouples were embedded in the center and both ends of the thickness of the epoxy resin.

The laminate was heated from a room temperature of 25°C to 180°C, and then air-cooled and water-cooled under different cooling conditions, water-cooled only on the epoxy side and water-cooled only on the aluminum side.

At this time, a thermocouple attached to the center of the epoxy resin was used to measure the temperature behavior that changed moment by moment.

Here, the heating method was as follows: after the laminate was placed in a furnace, the temperature was raised from room temperature to reach 180°C, and then the temperature was maintained for 10 minutes. As for the cooling method, air cooling was performed by turning off the power to the furnace from the heating and holding temperature, opening the door, and cooling the inside of the furnace.

On the other hand, the laminate was quickly put into water kept at room temperature to cool the laminate as a whole. Only the aluminum/epoxy resin side was water-cooled by installing a dedicated cooling device, and cooling was performed by spraying water only on one side. The amount of warp deformation that occurs at this time is defined as a residual amount of warp deformation δ_r.

On this occasion, the temperature is measured by a thermocouple attached to the center of the epoxy resin strip, and sometimes the warp deformation behavior of the laminated body that changes from moment to moment is photographed with a CCD camera, and the final warp amount of deformation was measured.

11.2.2 Physical Property Values Used for Analysis

Table 11.1 shows the thermal and mechanical properties of the constituent materials of the laminate used in the analysis.

Table 11.1 Properties of epoxy resin and printed board.

Materials \ Properties		Thermal conductivity λ (W/m K)	Specific heat C (kJ/kg K)	Density ρ (kg/m^3)	Glass transition temperature T_g (°C)	Thermal expansion coefficient α (10^{-6}/K)		Modulus E (GPa)
						$T \leqq T_g$	$T_g < T$	$T \leqq T_g$ ~ $T_g < T$
Epoxy resin	Rapid	0.186	1.423	1220	100.6	45.5	154.7	2.09 ~ 0.03
	Air				103.6	78.0	187.6	2.16 ~ 0.03
	Heat				92.9	95.2	158.0	1.79 ~ 0.02
Printed board		0.320	1.256	1810	119.2	14.5	12.2	9.73 ~ 1.80

Figure 11.2 shows the linear expansion coefficients of the epoxy resin and the printed circuit board, respectively.

Figure 11.3 shows the master curve of the relaxation modulus of both materials. The linear expansion coefficient α of the epoxy resin and the printed circuit board was calculated based on the measured value of thermal strain temperature change using a thermomechanical analyzer TMA-50.

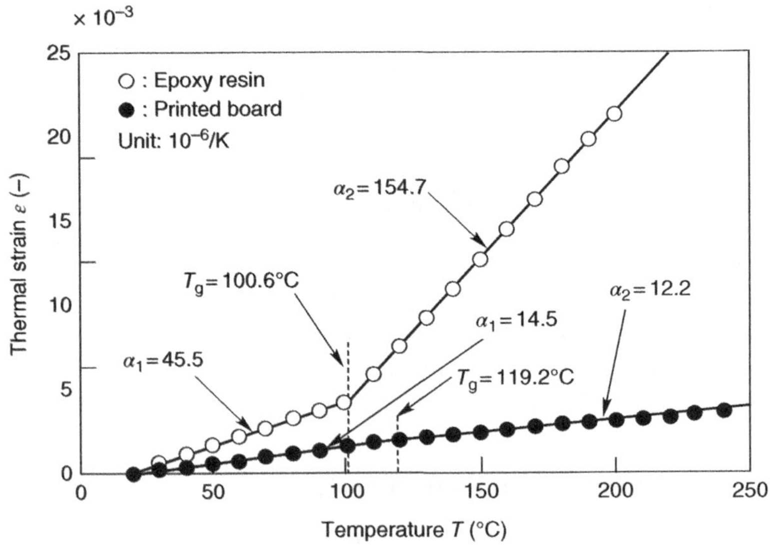

Figure 11.2 Thermal expansion coefficient for epoxy resin and printed board.

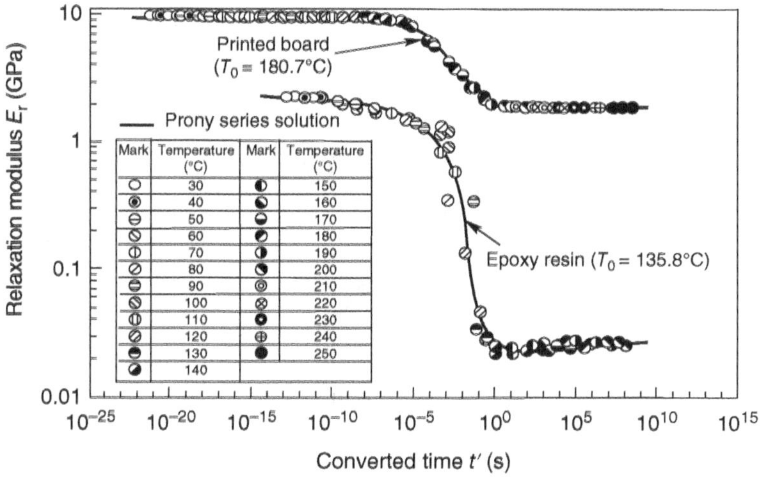

Figure 11.3 Master curve of epoxy resin and printed board.

In addition, the relaxation elastic modulus E_r is obtained by measuring the storage elastic modulus by changing the frequency of 1, 10, and 100 Hz at each desired temperature using a dynamic viscoelasticity measurement value Rheogel-E4000. These measurements are plotted on a converted logarithmic time axis under the reference temperature T_0.

As can be seen from Fig. 11.2, the T_g of the epoxy resin and the printed circuit board are 100.6°C and 119.2°C, respectively. This epoxy resin shows values of $\alpha_1 = 45.5 \times 10^{-6}$/K and $\alpha_2 = 154.7 \times 10^{-6}$/K, and the high-temperature part has larger values than the low-temperature part.

The printed circuit board shows $\alpha_1 = 14.5 \times 10^{-6}$/K and $\alpha_2 = 15.2 \times 10^{-6}$/K, indicating that the high-temperature part has lower values than the low-temperature part.

On the other hand, the relaxation modulus E_r of the epoxy resin and printed circuit board shown in Fig. 11.3 could be displayed with a single smooth master curve from an extremely short time to an extremely long time. Here, the relaxation modulus is approximated by the Prony series, while the activation energy ΔH in Fig. 11.4 approximates the time–temperature shift factor by the Arrhenius equation.

As can be seen from Fig. 11.4, the epoxy resin and the printed circuit board could be approximated by two Arrhenius equations with different activation energies ΔH at low and high temperatures.

From this, it can be seen that the epoxy resin and printed circuit board used here satisfy the time–temperature conversion rule and are materials to which the aforementioned linear viscoelastic theory can be applied, and these values are used in the VESAP analysis.

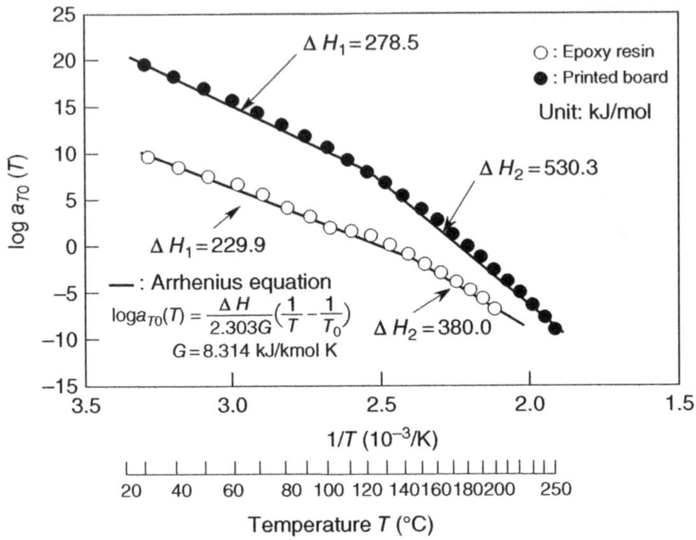

Figure 11.4 Time-temperature shift factor for epoxy resin and printed board.

11.2.3 Warp Deformation Analysis Conditions

A viscoelastic analysis software (VESAP) developed by the author based on the aforementioned linear viscoelastic theory was used for the viscoelastic analysis of warpage deformation.

The thickness of the epoxy resin was changed in three patterns as in the experiment, and a thermal analysis and a stress analysis were performed. The thermal and mechanical properties shown in Table 11.1 were used for the material properties used in the analysis.

The warp deformation analysis is the heating process from 25°C to 180°C and the cooling process from 180°C to 25°C. Also, in the cooling process, two cases of rapid cooling and air cooling were analyzed.

11.2.4 Results and Discussion

11.2.4.1 Heat Transfer Coefficient

The heat transfer coefficient was obtained by using the thermocouple attached to the center of the epoxy resin band plate described earlier, and correlating it with the experimental value of the temperature change over time.

As an example, Fig. 11.5 shows the results of the temporal change in the quenching temperature for an epoxy resin with a thickness of 5 mm.

Here, plots are experimental values, and solid lines are the results of viscoelastic numerical analysis. The heat transfer coefficient obtained from Fig. 11.5 was

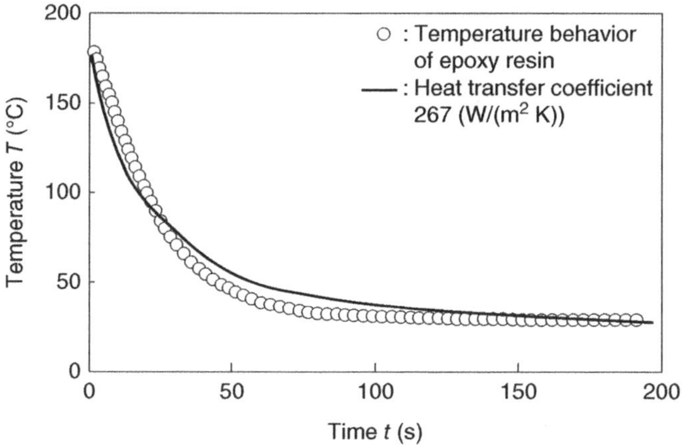

Figure 11.5 Temperature changes and heat transfer coefficient.

Table 11.2 Heat transfer coefficient of cooling condition.

Unit: (W/m^2 K)

Heat load condition \ Thickness of epoxy resin (mm)	2	5	8
Rapid	291	267	233
Air	10	10	9
Heat	15	15	14

267 W/(m^2 K). The heat transfer coefficients under other conditions were obtained by the same method, and their values are summarized in Table 11.2.

11.2.4.2 Warp Deformation Behavior

Figure 11.6a–c shows the warpage deformation behavior of various epoxy resin thicknesses when a series of thermal loads from heating to cooling are applied to the laminate.

That is, the experimental and theoretical values are shown for a series of heat loads during the heating process from room temperature of about 25°C to 180°C and the subsequent cooling process (air cooling, rapid cooling) from 180°C to 25°C.

The plots in Fig. 11.6 show the experimental values for air cooling and rapid cooling. The warpage deformation behavior during the heating process is shown on the left side of Fig. 11.6.

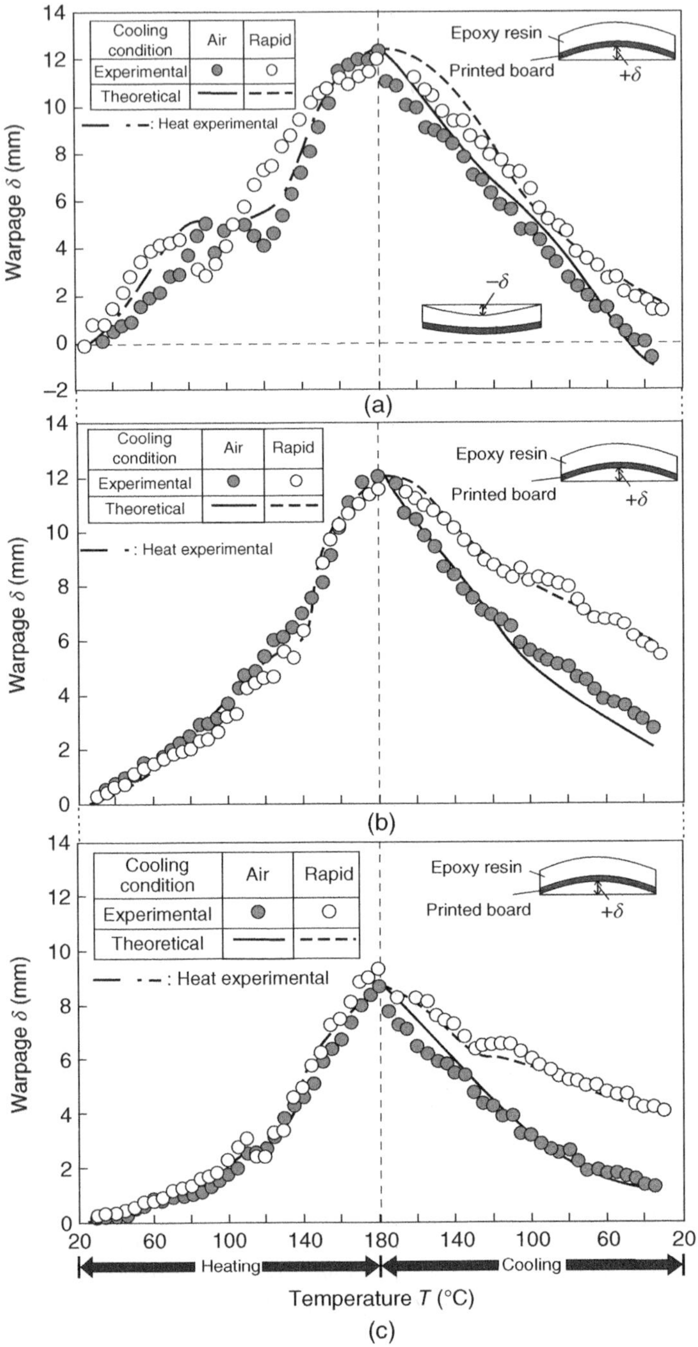

Figure 11.6 Warpage behavior of laminate by thermal load. (a) h_e = 2 mm, (b) h_e = 5 mm, (c) h_e = 8 mm.

Regarding the warpage deformation behavior of the laminated body, the epoxy resin side begins to warp with heating regardless of the epoxy resin thickness, and the amount of warpage deformation tends to increase as the temperature rises.

The amount of warpage changes in the temperature range from 80°C to 140°C, and then shows the maximum value at 180°C. It is thought that the elastic modulus and linear expansion coefficient of the epoxy resin and the printed circuit board, as well as the shape and dimensions, are greatly involved in the cause of such warping deformation behavior.

In the case of a laminate made of epoxy resin with a thickness of 2 mm, as shown in Fig. 11.6a, as the temperature rises, it begins to warp convexly toward the epoxy resin side when the gradient becomes gentle. This is thought to be due to the balance between the force of the printed circuit board trying to return to its original state and the thermal expansion force of the epoxy resin since the elastic modulus of the epoxy resin begins to decrease rapidly (2.09 GPa → 0.0267 GPa) near T_g.

Then, when the elastic modulus of the printed circuit board decreases (9.73 GPa → 1.80 GPa), it is thought that the thermal expansion force of the epoxy resin increases again.

On the other hand, in the case of Fig. 11.6b,c, the slope of the amount of warpage in the temperature range between 80°C and 140°C does not become gentle.

This is thought to be due to the fact that the thickness of the epoxy resin is thick, making it difficult for the printed circuit board to return to its original position.

The right side of Fig. 11.6a–c shows the warpage deformation behavior during the cooling process. In both cases, the amount of warpage deformation decreased due to the shrinkage of the epoxy resin upon cooling.

When air cooling and rapid cooling are compared here, air cooling reduces the amount of warpage deformation compared to rapid cooling. In addition, only in the case of rapid cooling of epoxy resin with a thickness of 8 mm, a change in the slope of the amount of warpage was observed at around 140°C.

Furthermore, as can be seen from these series of figures, the results of the viscoelastic numerical analysis agree well with the experimental values.

Therefore, the warpage deformation behavior can be predicted by viscoelastic analysis (VESAP) under both air cooling and quenching conditions.

11.2.4.3 Final Warp Deformation Amount and Residual Warp Deformation

Figure 11.7 shows the relationship between the epoxy resin thickness and the final warp deformation amount δ_f at the final stage (at 25°C) when a series of thermal loads were applied to the laminate from heating to cooling.

As can be seen from Fig. 11.7, the tendencies of the final warpage amount of rapid cooling and air cooling match.

In addition, when the epoxy resin thickness is 2 mm and air cooling is used, the printed circuit board side becomes convex, and a reverse warp phenomenon is observed.

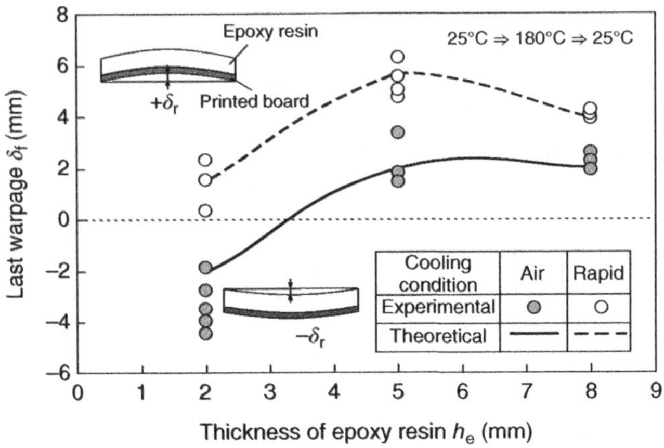

Figure 11.7 Last warpage of laminate against resin thickness.

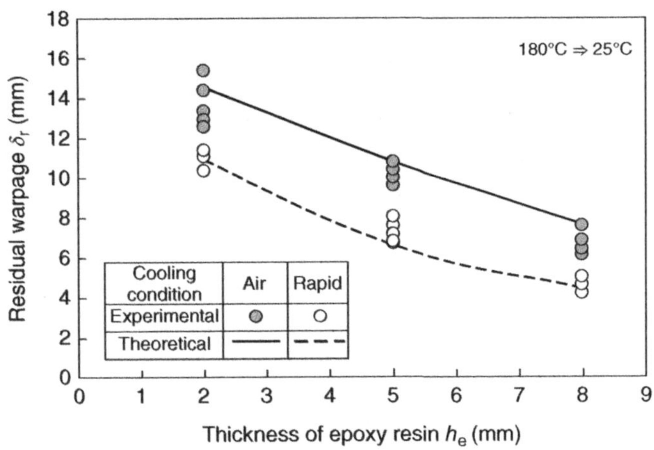

Figure 11.8 Residual warpage of laminate against resin thickness.

On the other hand, Fig. 11.8 shows the amount of warpage deformation δ_r (called the amount of residual warpage) and the epoxy resin thickness during cooling, assuming that the amount of warpage deformation is 0 at 180°C.

As can be seen from Fig. 11.8, the tendencies of the amount of residual warpage between rapid cooling and air cooling also match.

Furthermore, the amount of residual warpage is larger with air cooling at any epoxy resin thickness, and the difference in the amount of residual warpage between air cooling and quenching is almost constant.

However, the cause and mechanism of a large amount of residual warpage in the case of air cooling have not yet been elucidated. At present, we believe that this is due to the flexural rigidity of the laminate.

11.3 Two-Layer Laminate with Epoxy Resin/Steel

In this section, we discuss the warpage behavior of two-layer laminates consisting of epoxy resin and steel during the heating and cooling process when the thickness of each material is changed.

In this section, I will describe the shape of the laminate, the physical properties of the materials, and the comparison of the results of the experiment and the VESAP analysis on the warpage deformation behavior.

In addition, the comments are the same as those in Section 11.2 so that the reader can understand them, so the details are omitted.

1) Experimental Method (Figs. 11.9–11.13; Table 11.3)
2) Results of Experiments and analysis on warpage deformation behavior (Figs. 11.14–11.16)

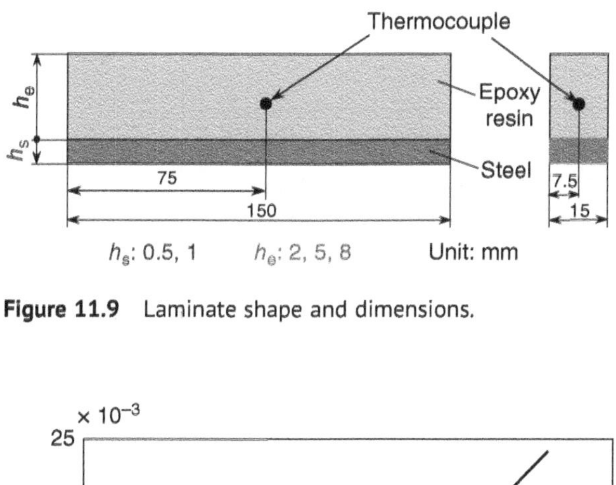

Figure 11.9 Laminate shape and dimensions.

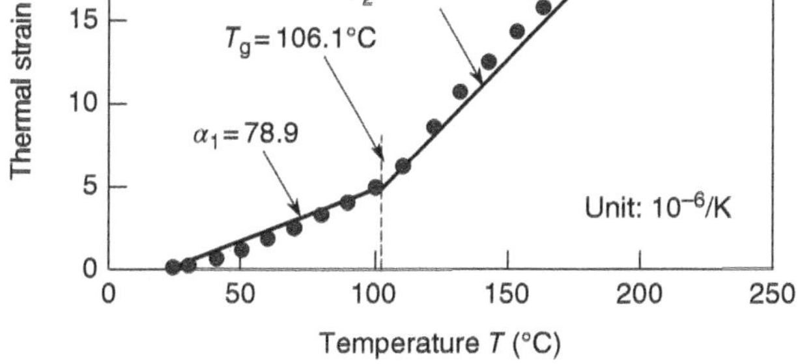

Figure 11.10 Thermal expansion coefficient for epoxy resin.

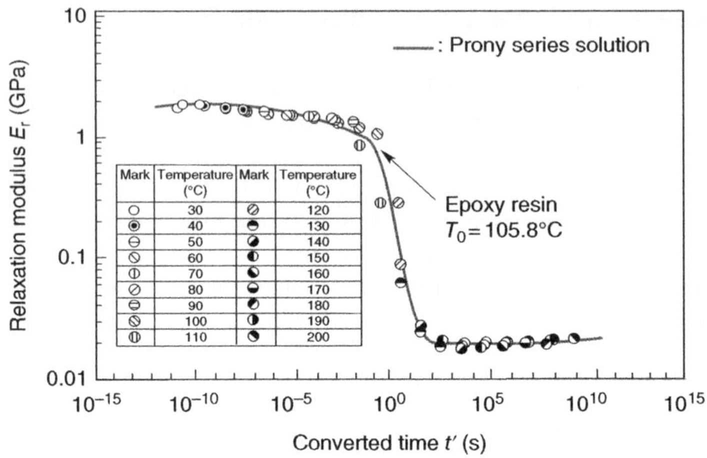

Figure 11.11 Master curve of epoxy resin.

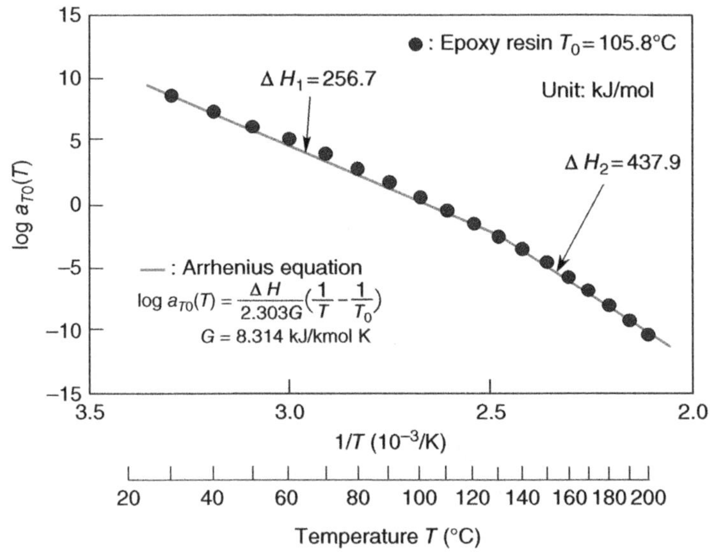

Figure 11.12 Time–temperature shift factor of epoxy resin.

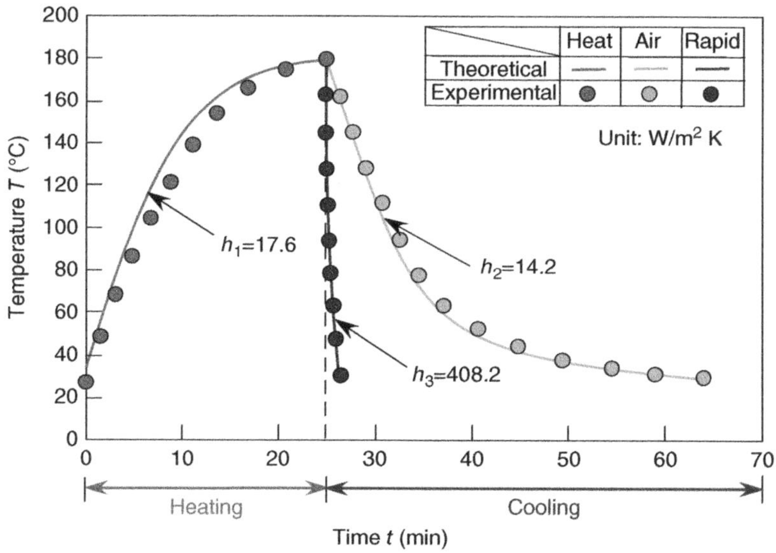

Figure 11.13 Temperature changes of laminate.

Table 11.3 Material properties of steel and epoxy resin.

Properties / Materials	Thermal conductivity λ (W/m K)	Specific heat C (kJ/kg K)	Density ρ (kg/m^3)	Glass transition temperature T_g (°C)	Thermal expansion coefficient α (10^{-6}/K)		Modulus E (GPa)	
					$T \leqq T_g$	$T_g < T$	$T \leqq T_g$	$T_g < T$
Steel	25.57	0.459	7750	—	10.4		200.0	
Epoxy resin	0.186	1.423	1220	106.1	78.9	227.9	1.79 ~ 0.02	

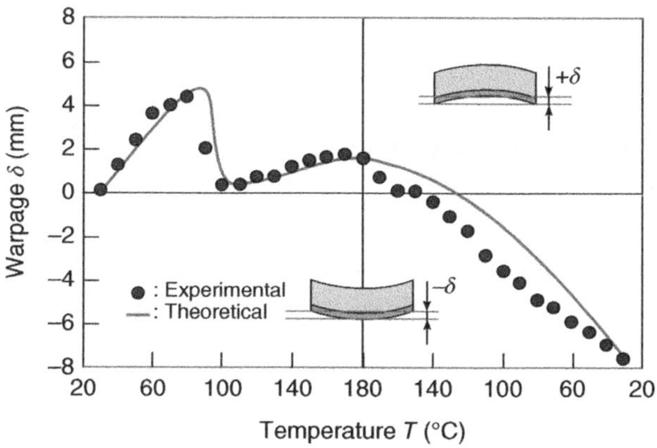

Figure 11.14 Warp deformation behavior of laminate from heating to air cooling (h_s: 0.5 mm, h_e: 2.0 mm).

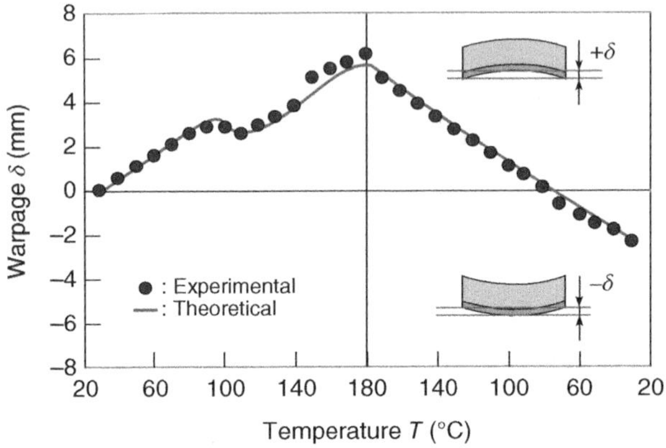

Figure 11.15 Warp deformation behavior of laminate from heating to air cooling (h_s: 0.5 mm, h_e: 5.0 mm).

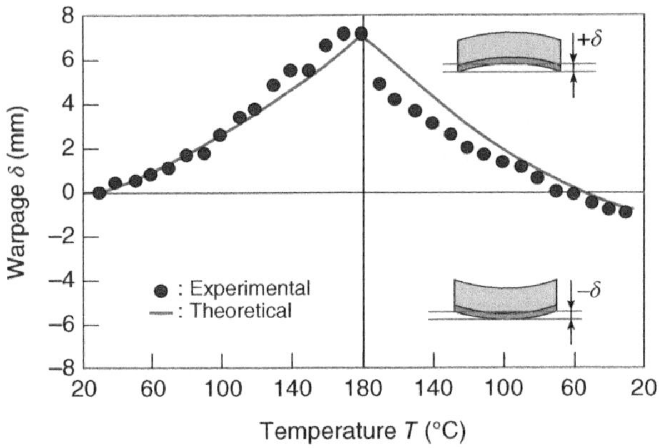

Figure 11.16 Warp deformation behavior of laminate from heating to air cooling (h_s: 0.5 mm, h_e: 8.0 mm).

11.4 Three-Layer Laminate with Steel/Epoxy Resin/Printed Board

In this section, I compare the experimental values of warpage deformation behavior and the results of VESAP analysis for three layers of steel/epoxy resin/printed board.

As with the example of the two-layer laminate in the previous section, a detailed description will be omitted.

1) Experimental method (Figs. 11.17–11.21; Table 11.4)
2) Results of experiments and analysis on warpage deformation behavior (Figs. 11.22–11.25)

Materials \ Pattern	I	II	III
Steel (h_s)	0.5	1.0	
Epoxy resin (h_e)	1.0	2.0	5.0
Printed board (h_p)	1.0	1.6	

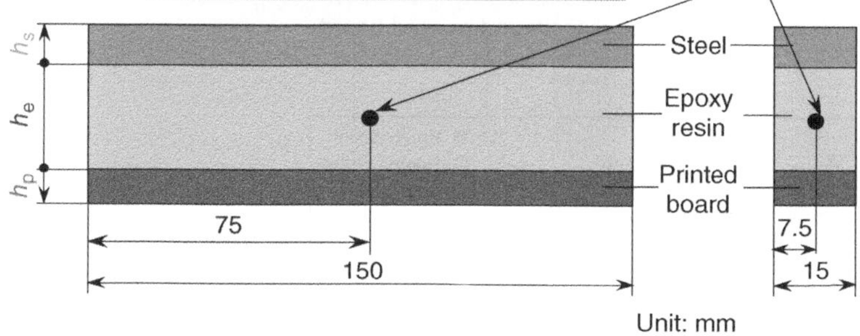

Figure 11.17 Shape and dimensions of three-layer laminate.

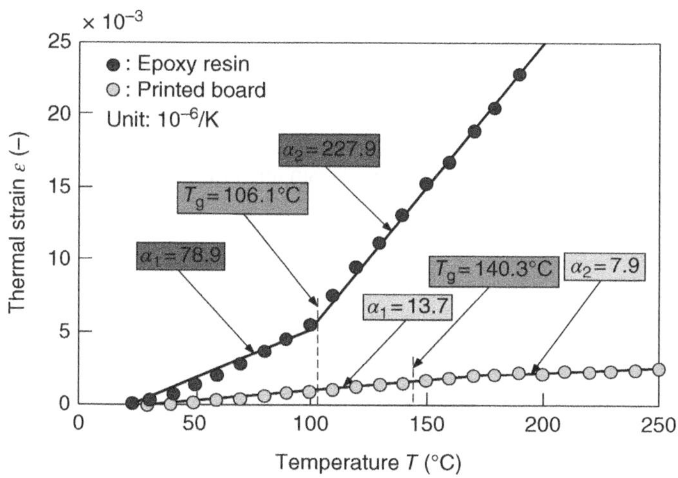

Figure 11.18 Thermal expansion coefficient for epoxy resin and printed board.

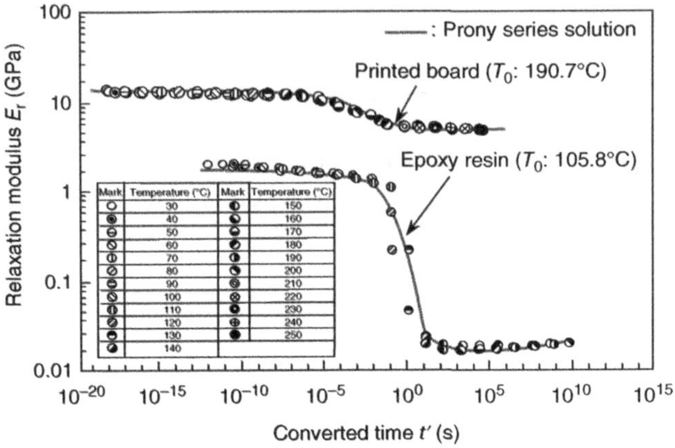

Figure 11.19 Master curve of epoxy resin and printed board.

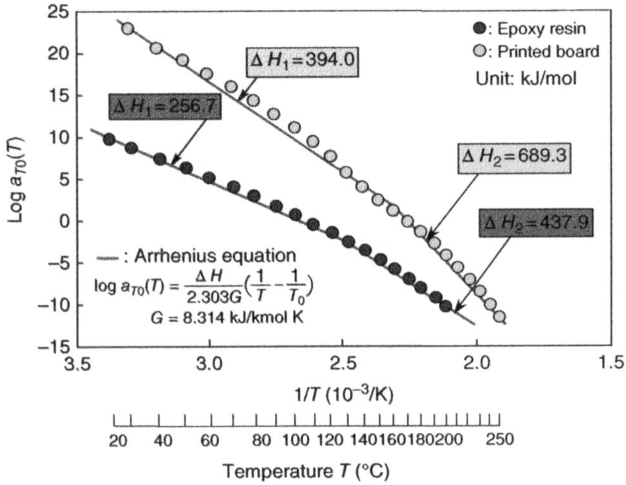

Figure 11.20 Time–temperature shift factor of epoxy resin and printed board.

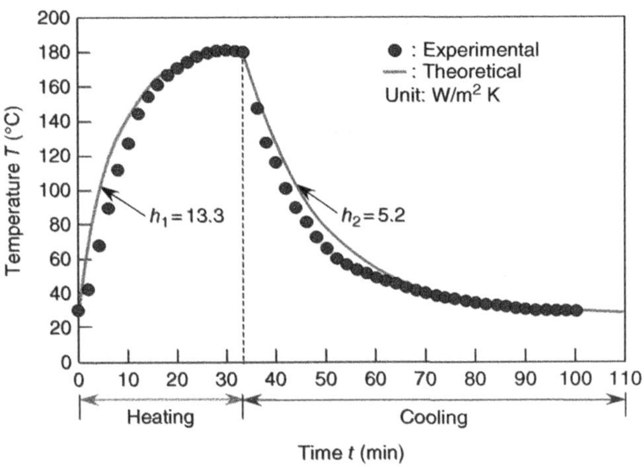

Figure 11.21 Temperature changes of laminate.

Table 11.4 Material properties of laminate.

Properties / Materials	Thermal conductivity λ (W/m K)	Specific heat C (kJ/kg K)	Density ρ (kg/m^3)	Glass transition temperature T_g (°C)	Thermal expansion coefficient α (10^{-6}/K)		Modulus E (GPa)	
					$T \leqq T_g$	$T_g < T$	$T \leqq T_g$	$T_g < T$
Steel	25.57	0.459	7750	—	10.4		200.0	
Epoxy resin	0.186	1.423	1220	106.1	78.9	227.9	1.79 ~ 0.02	
Printed board	0.33	1.256	1810	138.8	12.6	6.7	9.71 ~ 4.84	

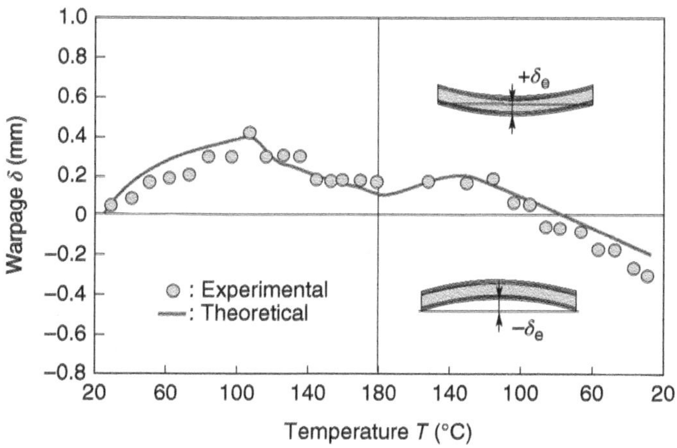

Figure 11.22 Warp deformation behavior of laminate from heating to cooling (h_s: 1 mm, h_e: 1 mm, h_p: 1.6 mm).

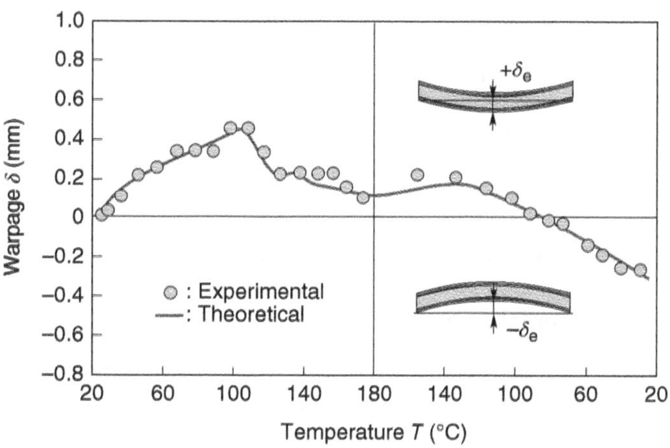

Figure 11.23 Warp deformation behavior of laminate from heating to cooling (h_s: 1 mm, h_e: 2 mm, h_p: 1.6 mm).

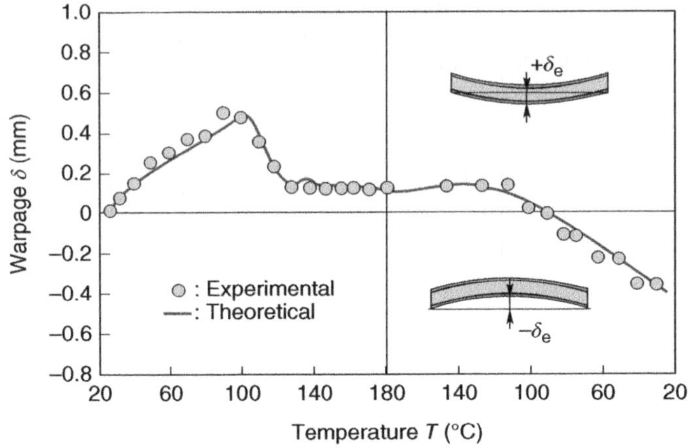

Figure 11.24 Warp deformation behavior of laminate from heating to cooling (h_s: 1 mm, h_e: 5 mm, h_p: 1.6 mm).

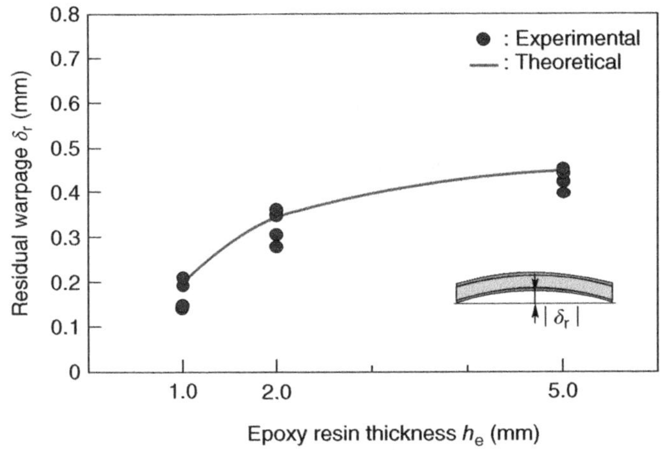

Figure 11.25 Amount of residual warpage deformation at epoxy resin thickness (h_s: 1.0 mm, h_p: 1.6 mm).

11.5 Analysis Experiment of Four-Layer Laminate

11.5.1 Method of Experimental Analysis

This chapter targets a laminated body composed of four layers, which is more complicated than Section 11.4. The laminate was epoxy resin/steel and Al alloy/epoxy resin/printed board.

Here, the epoxy resin and the printed circuit board were fixed, and the internal material was a sandwich structure of steel and Al alloy.

Then, as in Section 11.4, the warp deformation behavior that occurs when a series of heat loads from heating to cooling is applied to the laminate was examined in both VESAP analysis and experiment.

Specifically, we focused on the relationship between the thickness of the constituent materials and the viscoelastic properties of the epoxy resin and printed circuit board, which affect the warp deformation behavior.

Figure 11.26 shows the shape and dimensions of the four-layer laminate used in the analysis and experiment. The constituent materials used were the same two-component mixed resin, SUS420J2 for the steel material, A1050P for the Al alloy and the FR-4 substrate for the printed circuit board. The dimensions of the laminate are 150 mm in length, 15 mm in width, 0.5 mm in thickness for steel, 1 mm for aluminum alloy, and 1 mm for printed substrate, and the center is the upper epoxy resin (hereinafter abbreviated as U epoxy).

For both the partial epoxy resins (hereinafter abbreviated as I epoxy), three types of 1, 2, and 5 mm were used, and a total of 18 combinations of laminates were used. The method for manufacturing the laminated body is the same as described earlier.

Materials \ Pattern		I	II	III
1	Epoxy resin (h_{e1})	1.0	2.0	5.0
2	Steel (h_s)	0.5		
3	Aluminum (h_a)	1.0		
4	Epoxy resin (h_{e2})	1.0	2.0	5.0
5	Printed board (h_p)	1.0		

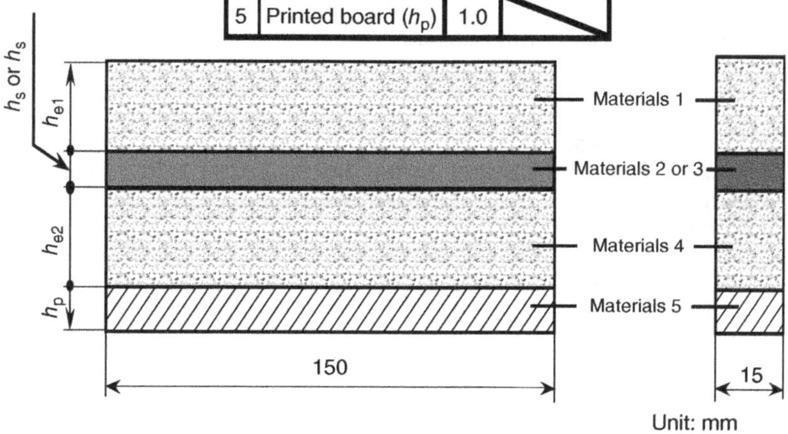

Figure 11.26 Shape and dimension of four-layer laminate.

The temperature and deformation of the laminate are measured in the same way as in the two-layer and three-layer measurements. For the heat load, heat from about 25°C to 180°C under the condition of 3°C/min and then cool to room temperature in the dryer.

At that time, the changing temperature was measured, and the warp deformation behavior of the laminated body was sequentially photographed with a CCD camera, and the image was taken into a personal computer to measure the warp deformation amount.

11.5.2 Physical Characteristics of Constituent Materials

Table 11.5 shows the material property values used in the VESAP analysis. Figure 11.27 shows the temperature dependence of the thermal strain of the epoxy resin and the printed circuit board. The glass transition temperatures T_g of the epoxy resin and the printed circuit board are 104°C and 139°C, respectively, and the coefficient of linear expansion is 90 and 14 ppm/K in the low-temperature part α_1 (T_g or less), respectively, and the high-temperature part α_2 (T_g or more) at 223 and 8 ppm/K, respectively.

Figure 11.28 shows the master curve of the relaxation elastic modulus E_r of the epoxy resin and the printed circuit board. From Fig. 11.28, E_r can be represented by a single smooth master curve from a very short time to a very long time. For viscoelasticity analysis, a value that approximates this to the Prony series is used.

In addition, Fig. 11.29 shows the time–temperature shift factor when creating the master curve of both materials. As can be seen from Fig. 11.29, the activation energy ΔH differs between the low-temperature part and the high-temperature part in both materials, and in the viscoelastic analysis by VESAP, these are

Table 11.5 Physical properties of constituent materials.

Properties / Materials	Thermal conductivity λ (W/m K)	Specific heat C (kJ/kg K)	Density ρ (kg/m^3)	Glass transition temperature T_g (°C)	Thermal expansion coefficient α (10^{-6}/K)		Modulus E (GPa)	
					$T \leqq T_g$	$T_g < T$	$T \leqq T_g$	$T_g < T$
Epoxy resin	0.19	1.42	1220	104	90	223	2.1 ~ 0.02	
Printed board	0.33	1.26	1810	139	13	7	9.7 ~ 4.9	
Steel	25.6	0.46	7750	—	10		200	
Aluminum	228	0.91	2700	—	24		68	

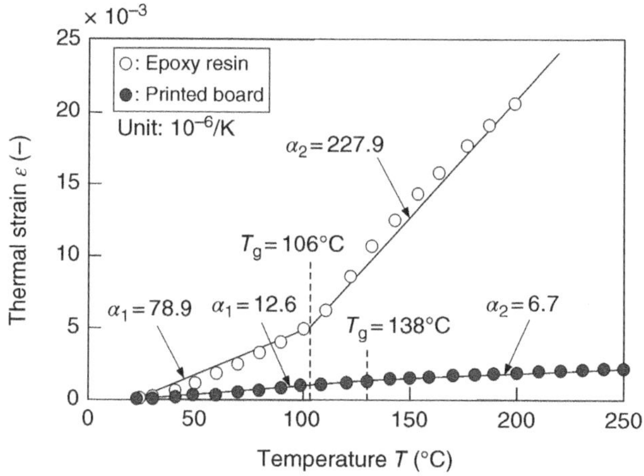

Figure 11.27 Thermal strain for epoxy resin and printed board.

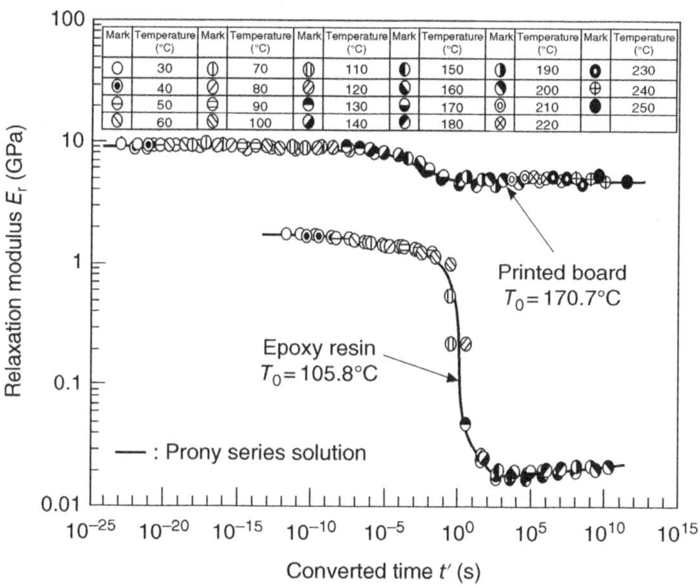

Figure 11.28 Relaxation modulus for epoxy resin and printed board.

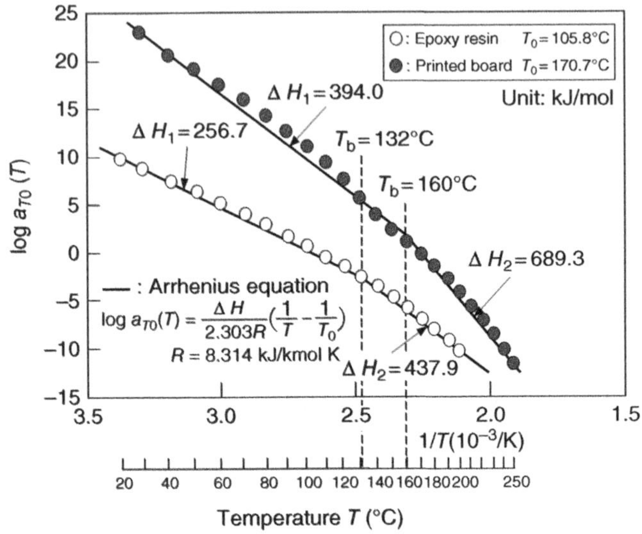

Figure 11.29 Time–temperature shift factor for epoxy resin and printed board.

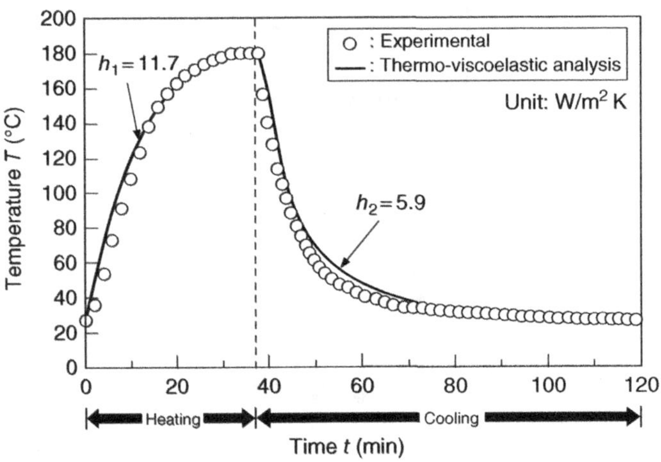

Figure 11.30 Temperature changes of heating and cooling.

approximated by two Arrhenius equations, and the amount of warpage deformation is shown in Fig. 11.29.

Figure 11.30 shows the temperature change from heating to cooling of the laminate. Here, the plot is the measured value by the thermocouple embedded in the laminated body, and the heat transfer coefficient h calculated by the thermal analysis corresponding to the measured value is 11.7 W/(m^2 K) at the time of heating and 5.9 W at the time of cooling.

In general, the heat transfer coefficient indicates the degree of heat dissipation from the surface of an object, and if the object has irregularities, the value will differ depending on the location, but in this experiment, since it is a smooth laminate, thermal analysis by VESAP and thermal stress. This value was used for the coupled analysis of the deformation analysis.

11.5.3 Results and Considerations

11.5.3.1 Warp Deformation Behavior of Laminates Containing Steel

Figure 11.31a–c show the warp deformation behavior when a series of heat loads are applied to the laminate, comparing the VESAP analysis values and the experimental values.

Here, the thickness dimensions of the laminated body are constant at 0.5 and 1 mm for the steel material and the printed circuit board, respectively. The case of U epoxy 1 mm and I epoxy 1 mm, and the case of Fig. 11.31c shows the case of U epoxy 1 mm and I epoxy 5 mm.

Here, the plot is the experimental value, and the solid line is the VESAP analysis value. In the thickness configuration of the laminate shown in Fig. 11.31a, the U epoxy side begins to warp in a convex state during the heating process, but then decreases once and then increases again toward flattening, and at 180°C, it is 1 mm. The degree of warpage deformation is shown in the cooling process from 180°C; it gradually decreased and became flat near 100°C, and at 25°C after cooling, the U epoxy side turned into a concave state and showed complicated and interesting deformation behavior.

In the thickness configuration of the laminated body shown in Fig. 11.31b, since the thickness of both U epoxy and I epoxy is as thin as 1 mm, the warp deformation of the laminated body is almost the same with respect to the heat load from heating to cooling.

This is because the elastic modulus of the steel material and the printed substrate is extremely large. So, the bending rigidity of the laminate is also large, and even if the thermal expansion of U epoxy and I epoxy is large, the elastic modulus is small. So, the laminate is only deformed.

It is probable that the bending moment of the laminated body did not act, and as a result, the amount of warp deformation of the laminated body became smaller due to the balance of the forces of the constituent materials.

On the other hand, when the U epoxy in Fig. 11.31c is thin and the I epoxy is thick, the tendency of warpage is opposite to that in Fig. 11.31a, but warpage deformation is hardly observed as in Fig. 11.31b.

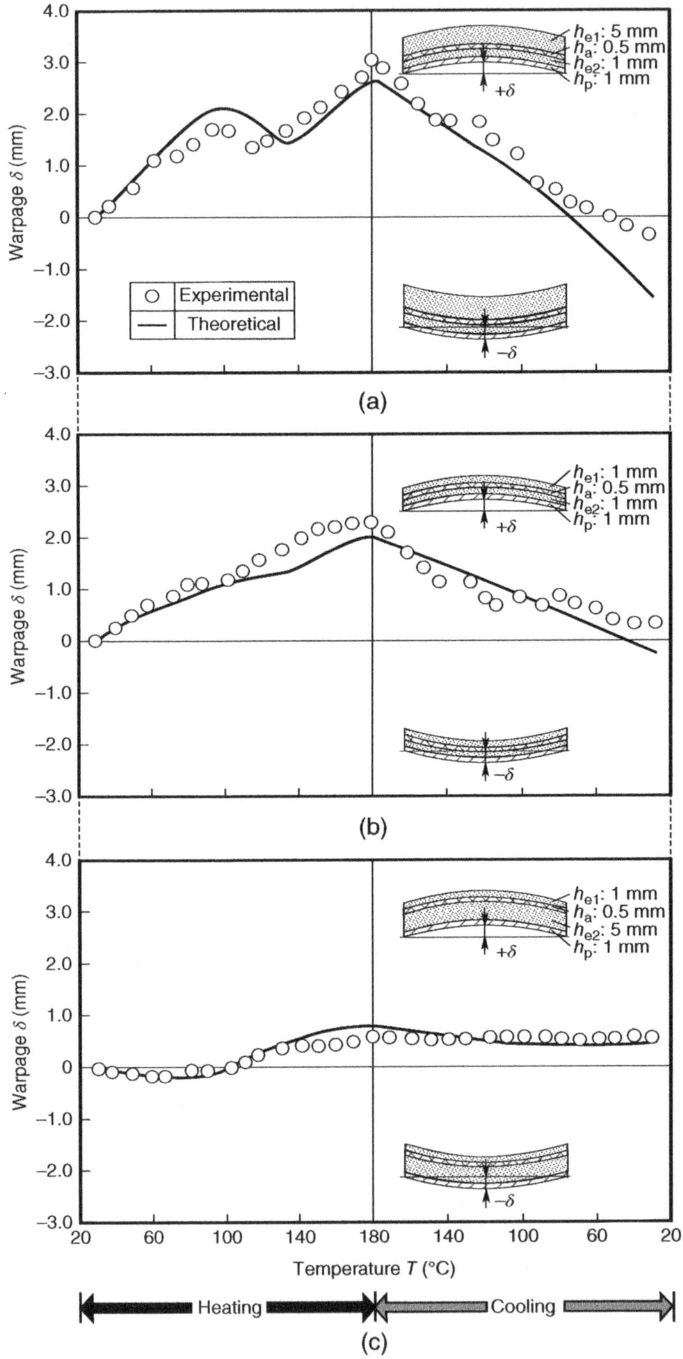

Figure 11.31 Warp deformation behavior from heating to cooling (thickness of steel is 0.5 mm). (a) h_{e1}: 5 mm, h_{e2}: 1 mm; (b) h_{e1}: 1 mm, h_{e2}: 1 mm; (c) h_{e1}: 1 mm, h_{e2}: 5 mm.

In any case, the VESAP analysis value reflects the experimental value, and it can be seen that it is an effective tool.

In a laminate composed of such three types of materials, the thickness of the epoxy resin, steel material, and printed substrate and their viscoelastic property values are mutually exclusive with the flexural rigidity of the entire laminate and the thermal expansion/contraction force of each material.

Moreover, it is presumed that they influence each other in a complicated manner to determine the final amount of warp deformation of the laminated body.

11.5.3.2 Warp Deformation Behavior of Laminate Containing Al Alloy

Figure 11.32a–c shows the warp deformation behavior that occurs when a heat load from heating to cooling is applied to a four-layer laminate containing an Al alloy and a resin.

In Fig. 11.32, the thickness of the aluminum alloy and the printed circuit board are both constant at 1 mm, and the combination of epoxy resin is the same as in the previous section.

From Fig. 11.32a, the warp deformation increases with heating in a convex state on the U epoxy side, decreases at the glass transition temperature of the epoxy resin near 100°C, but increases thereafter, and is 3 mm at the maximum temperature of 180°C.

This value is larger than that of the steel material in the previous section. In addition, the amount of warpage deformation temporarily decreases near 100°C, which is due to the decrease in the elastic modulus of the glass transition temperature of the epoxy resin, as in the case of steel materials.

Next, in the cooling process, the heat shrinkage force of the thick U epoxy becomes dominant as in the case of steel materials, and the laminate tends to flatten, and when further cooled, it exhibits a reverse warpage phenomenon and U at the end temperature of 25°C.

The epoxy side becomes concave, and the value is about 0.5 to 1 mm, which is slightly smaller than that of steel.

On the other hand, in the case of the laminated body having the dimensional configuration shown in Fig. 11.32b, the warp deformation behavior was almost the same as that in Fig. 11.32a, and the value was about 2 mm at 180°C.

In the cooling process, it decreases monotonically and becomes almost flat at 25°C. Here, it can be seen that in the case of the steel material, almost no warp deformation occurred, whereas, in the case of the Al alloy, the amount of warp deformation was also large.

This difference will be done due to the difference in elastic modulus and linear expansion coefficient between steel and aluminum alloy, and is determined by the relationship between the bending moment due to the thermal expansion and contraction force of the constituent materials including epoxy resin and the bending rigidity of the laminate.

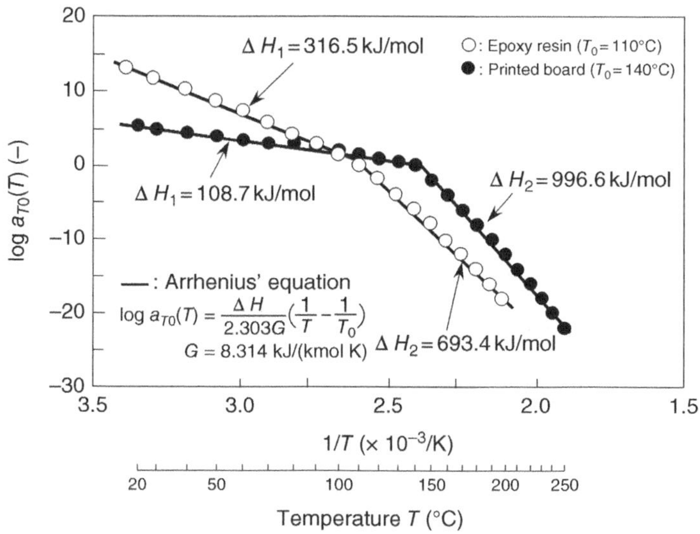

Figure 12.4 Time-temperature shift factor.

Here, for the viscoelastic analysis, the master curve of E_r in Fig. 12.3 was approximated by the Prony series, and the activation energy ΔH in Fig. 12.4 was approximated by the Arrhenius equation.

In the analysis, it was assumed that the heat generation temperature of the epoxy resin was the maximum value, and at this point, the curing reaction was completed and the epoxy resin became a solid state. Therefore, using the physical characteristics of the solid epoxy resin, the analysis was performed as cooling from the maximum heat generation temperature of various epoxy resin thicknesses to room temperature of 25°C.

12.4 Simple Prediction Formula for Calculating Curing Warpage

12.4.1 Guidance of Simple Prediction Formula

Figure 12.5 shows the shape of the laminate consisting of material 1 and material 2 used in the simple prediction formula. Let L be the length, b be the width, and h_1 and h_2 be the thicknesses of material 1 and material 2, respectively, and the distance ξ from the lowest end of material 2 to the neutral axis N–N can be expressed by Eq. (12.1).

Here, the distances from the center of material 1 and material 2 to the neutral axis are h_1' and h_2', respectively. When a heat load is applied to this laminated

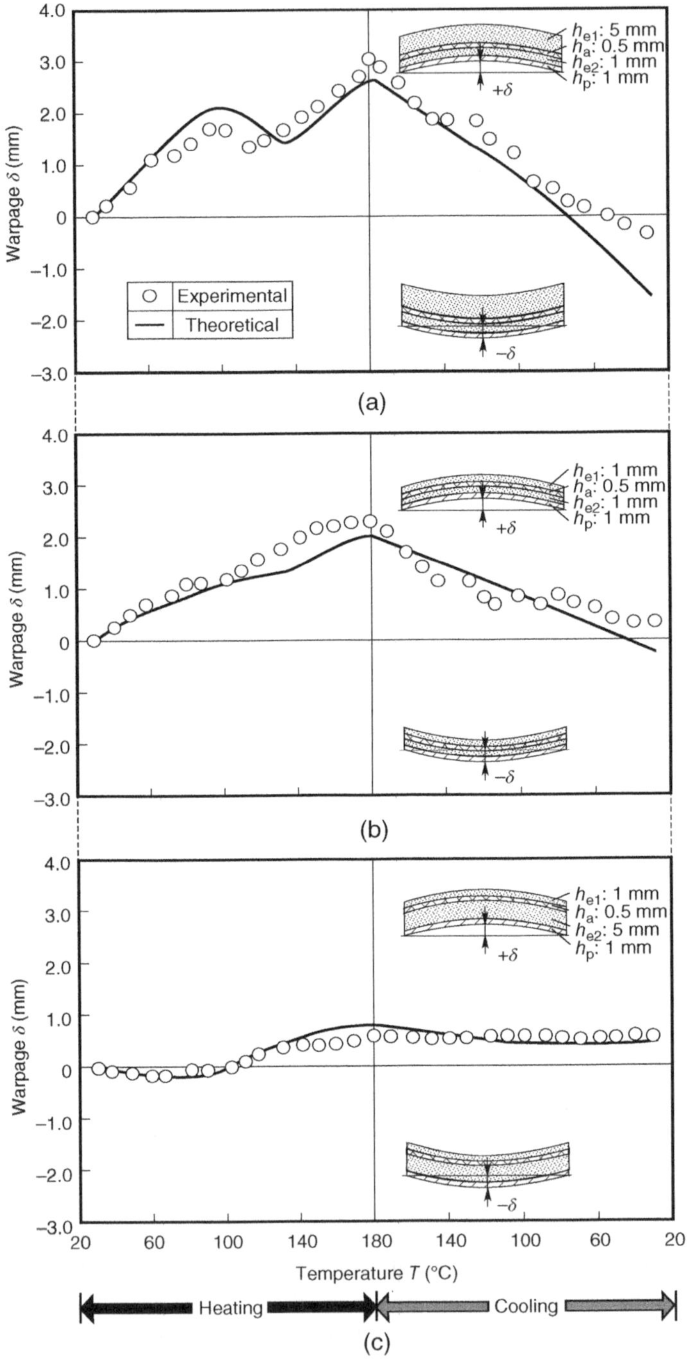

Figure 11.32 Warp deformation behavior from heating to cooling (thickness of Al alloy is 1 mm). (a) h_{e1}: 5 mm, h_{e2}: 1 mm; (b) h_{e1}: 1 mm, h_{e2}: 1 mm; (c) h_{e1}: 1 mm, h_{e2}: 5 mm.

In the case of the laminated body having the dimensional configuration shown in Fig. 11.32c, unlike Fig. 11.32a,b, the U epoxy side becomes slightly concave, and then becomes convex when the temperature rises.

The amount of warpage deformation is small even in the cooling process. The amount of warp deformation at 25°C also remained in a slightly convex state on the U epoxy side, and this result is in good agreement between the VESAP analysis value and the experimental value.

11.5.3.3 Key Points of Warp Deformation Behavior of Laminates Made of Different Materials

As described above, a complex four-layer laminate that combines an elastic steel material or an Al alloy with an epoxy resin or printed circuit board that exhibits remarkable viscoelastic behavior is warped and deformed by applying a series of heat loads from heating to cooling. The behavior was examined by VESAP analysis and experiment.

As a result, it was clarified that the warp deformation behavior of the laminate made of such a viscoelastic material is complicated and exhibits an interesting phenomenon. That is, the manufacturing process in which various viscoelastic property values such as the elastic coefficient and linear expansion coefficient of the constituent material, which are greatly affected by temperature and time, and the glass transition temperature, and the bending rigidity of the constituent are exposed to a high temperature of about 260°C or higher.

In addition, it is expected that they will affect each other in a complicated manner with the thermal load applied in various reliability tests before shipment, causing warpage deformation, peeling of the interface, and defects such as resin cracks.

In particular, since the elastic coefficient and linear expansion coefficient of resin change significantly near the glass transition temperature, thermal strain is applied in combination with a material such as an LSI chip that exhibits elastic behavior depending on the temperature conditions and temperature range in which it is loaded. And thermal stress will change significantly, resulting in interfacial peeling and warpage deformation.

Furthermore, since resin is a material having characteristic properties such as creep phenomenon and stress relaxation phenomenon whose properties change greatly with temperature and time, it is more difficult to elucidate the cause of such defects and the mechanism of occurrence of such defects.

Therefore, it is becoming more important to be familiar with the viscoelastic properties of resins and the like and to optimize the overall design, especially the thermal process during manufacturing, which greatly affects warpage deformation.

For this reason, viscoelasticity simulation technology is even more indispensable for obtaining efficient design guidelines for the constituent materials and dimensions of various electronic components used in electronic devices, heat load, and shortening the development period.

12

Deformation Prediction Method Considering Curing Shrinkage of Resin

12.1 Introduction

Up to Chapter 11, I have been discussing the warp deformation problem that occurs in a series of heat load processes from heating to cooling, which acts during manufacturing, for various laminates containing resin, was investigated from both aspects of Visco-Elastic Stress Analysis Program (VESAP) analysis and experiments considering viscoelastic behavior.

In this chapter, I have focused on the curing shrinkage, which is an essential property of resins. In other words, we focused on the effect of curing shrinkage on the warp deformation behavior of the laminate in response to the curing process due to the chemical reaction of the epoxy resin and examined the relationship with the thermal expansion and contraction in Chapters 8–11.

In general, epoxy resin is a two-component type, and cross-linking proceeds by a chemical/curing reaction, and the state changes from a liquid to a solid through gelation. The epoxy resin that the author has discussed in the VESAP analysis has been a general-purpose epoxy resin that involves this chemical reaction.

Therefore, many engineers who handle resins that enclose large-scale integration (LSI) chips are affected by the curing shrinkage of the resin, in addition to the warp deformation and thermal stress of electronic devices caused by thermal expansion and contraction due to the thermal load of solid materials. It seems to be a problem.

The same is true for the author, but I have no chemical knowledge and ask an expert for details. In this chapter, I will explain the basic experiments and the concept of simple theory with curing shrinkage as a problem.

Polymeric Materials for Electronic Packaging, First Edition. Shozo Nakamura.

Published 2023 by John Wiley & Sons, Inc.

12.2 Examination the Procedure and the Way of Thinking

Figures 12.1 and 12.2 show the concept and procedure of the research. Figure 12.1 schematically shows the exothermic temperature and the warped deformation state when the epoxy resin changes from a liquid to a solid through gelation with a curing reaction.

As shown in Fig. 12.1, the liquid phase is up to time t_1, the gel phase is from t_1 to t_2, and the solid phase is after t_2.

In addition, the exothermic temperature T rises with the curing reaction, reaches a maximum in the solid phase at time t_2, and then falls with the passage of time.

On the other hand, the experimental value δ_{ex} of the total warp deformation amount of the laminated body is the warp deformation amount due to the curing reaction of the epoxy resin (hereinafter abbreviated as cured warp δ_{cu}) and the warp deformation amount due to the temperature change in the solid state (hereinafter abbreviated as thermal warp δ_{th}) can be expressed as the sum.

It is considered that the warp deformation of the laminated body starts from the gelation start time t_1 of the epoxy resin and then increases with the passage of time to become a solid at the time t_2 when the exothermic temperature reaches the maximum value T_{max}.

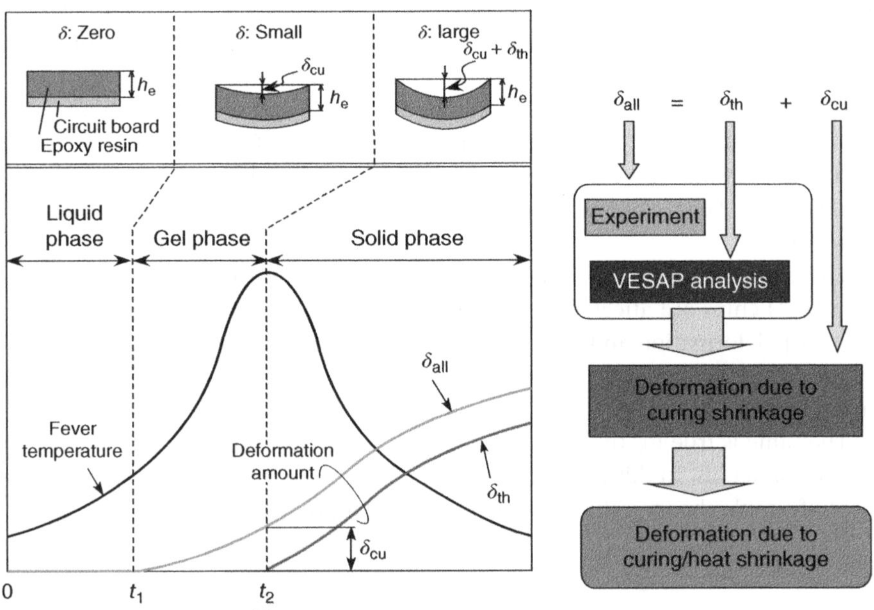

Figure 12.1 Concept of examination considering curing shrinkage.

Since it is safe to treat the epoxy resin as a solid state thermodynamically after time t_2, it is considered that the thermal warp δ_{th} starts from this point. Then, the difference between δ_{ex} and δ_{th} at time t_2 can be considered the curing warp δ_{cu} caused by the curing reaction.

That is, the total warp deformation amount δ_{tc} of the laminated body in the final state t_∞ after sufficient time has passed can be expressed as

$$\delta_{tc} = \delta_{cu} + \delta_{th} \tag{12.1}$$

It has been confirmed that such an idea is generally valid from past data and preliminary experiments with strain gauges. In addition, the warp deformation amount δ is naturally a function of temperature.

Figure 12.2 shows the examination procedure. Theoretical values of δ_{cu} and δ_{th} are obtained independently, and their validity is compared with the experimental values.

Specifically, the thermal warpage δ_{th} is obtained by VESAP analysis, while the curing warpage δ_{cu} is calculated from the simple prediction formula described later based on the actual measurement of the curing behavior of the epoxy resin.

Then, by comparing these with the total warp deformation amount δ_{ex} obtained in the experiment, a simple method for predicting the warp deformation amount that progresses in both the curing reaction and the temperature change was formulated.

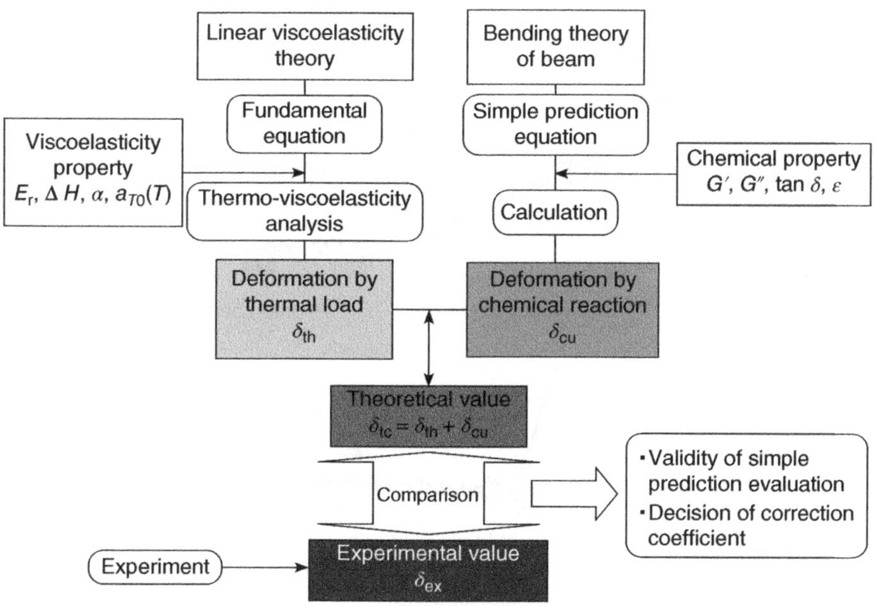

Figure 12.2 Flow of study.

12.3 Contents of VESAP Analysis

For the viscoelasticity analysis, the aforementioned basic formula derived based on the linear viscoelasticity theory was used as before. Table 12.1 shows the viscoelastic property values of the constituent materials used in the analysis.

In addition, Figs. 12.3 and 12.4 show the master curve of the relaxation modulus E_r of the epoxy resin and the printed circuit board, and the time–temperature shift factor $a_{T0}(T)$ when creating the master curve, respectively.

Table 12.1 Material properties of the laminate.

Property / Material	Thermal conductivity λ (W/m K)	Specific heat C (kJ/kg K)	Density ρ (kg/m^3)	Glass transition temperature T_g (°C)	Elastic modulus E (GPa)	Expansion coefficient α (10^{-6}/K)	
						$T \leqq T_g$	$T_g < T$
Epoxy resin	0.186	1.423	1220	104	2.1–0.02	90	227.9
Printed board	0.372	1.260	1810	139	9.7–4.9	126	6.7
Steel	14.7	0.452	7870	—	197	11.7	

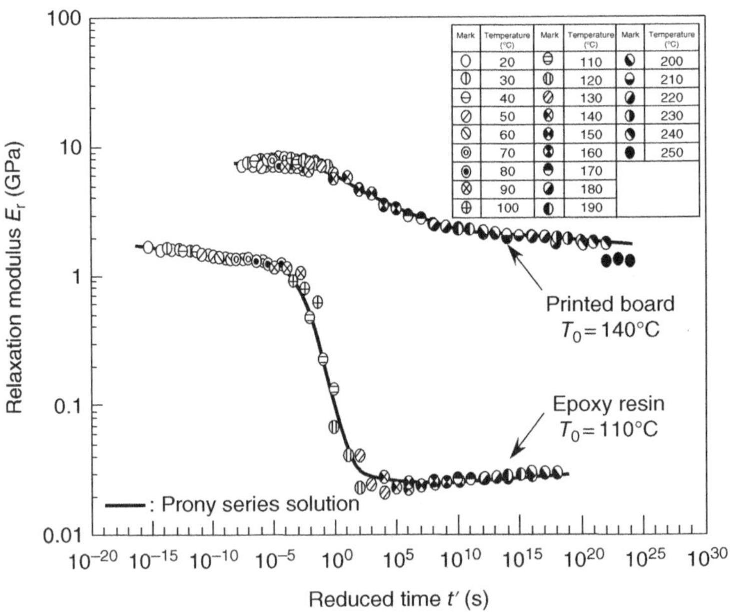

Figure 12.3 Master curve of relaxation modulus.

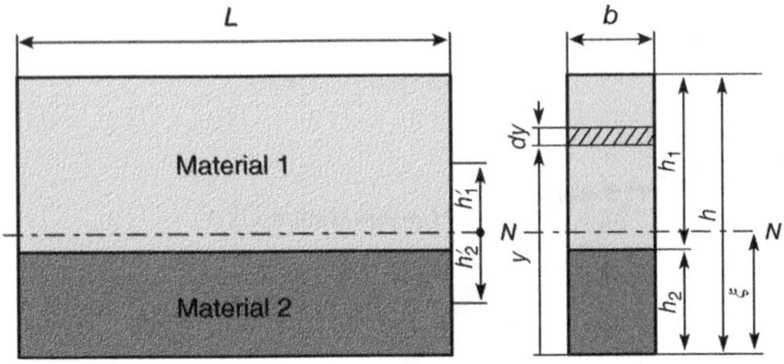

Figure 12.5 Shape and dimensions of laminated body for simplified solution.

body, a thermal expansion force and contraction force are generated in each material, and it is considered that the laminated body is warped and deformed depending on the balanced state of this force.

E_1 and E_2 are Young's modulus of material 1 and material 2, respectively, A_1 is the cross-sectional area of material 1 (bh_1), and A_2 is the cross-sectional area of material 2 (bh_2).

Further, assuming that the moment of inertia of area with respect to the neutral axis of material 1 and material 2 is I_1 and I_2, respectively, I_1 and I_2 can be expressed by the following Eqs. (12.2) and (12.3).

The moment of inertia of area is the basis of strength of materials, and can be expressed by dimensions such as the thickness, width, and diameter of the object, and is a physical quantity that is an index of whether the object is easily deformed by an external force, that is, the object.

It is an index showing rigidity.

$$\xi = \frac{E_1 \int_{A_1} y\,dA_1 + E_2 \int_{A2} y\,dA_2}{(E_1 A_1 + E_2 A_2)} \tag{12.2}$$

$$I_1 = \frac{\{h_1^3 - 3(\xi - h_2) \cdot h_1^2 + 3(\xi - h_2)^2 \cdot h_1\} \cdot b}{3} \tag{12.3}$$

$$I_2 = \frac{(h_2^3 - 3\xi h_2^2 + 3\xi^2 h_2) \cdot b}{3} \tag{12.4}$$

Now, the thermal strain ε_m generated in the material m (hereinafter, m represents the material 1 and the material 2) when a heat load is applied is expressed by the following equation, where the linear expansion coefficient α_m and the temperature difference ΔT_j of the heat load are used.

$$\varepsilon_m = \alpha_m \Delta T_j, \quad m = 1, 2, \ j = 1 \sim n \tag{12.5}$$

Assuming that the thermal stress σ_m generated in the material m causes minute strain in this laminate, E_m is the Young's modulus of the material m according to the Hooke's law.

$$\sigma_m = \varepsilon_m E_m, \quad m = 1, 2 \tag{12.6}$$

The heat shrinkage force P_m generated in the material m is represented by

$$P_m = \sigma_m A_m, \quad m = 1, 2 \tag{12.7}$$

The bending moment M_m generated in the material m is the product of the heat shrinkage force P_m and the distance $h'_m$from the center of the cross section of the material m to the neutral axis.

$$M_m = P_m \cdot h'_m, \quad m = 1, 2 \tag{12.8}$$

Assuming that the warp deformation of this laminate is caused by the bending moments M_1 and M_2 acting on material 1 and material 2, the bending moment M_d involved in the warp deformation is set to

$$M_d = P_1 \cdot h'_1 - P_2 \cdot h'_2 \tag{12.9}$$

The radius of curvature of the laminate is determined from the conventional bending theory.

$$R = \frac{(E_1 \cdot I_1 + E_2 \cdot I_2)}{M_d} \tag{12.10}$$

The increment of the amount of deformation corresponding to the temperature change $\Delta\delta_j$ is given in

$$\Delta\delta_j = \frac{L}{2} \tan\left\{\frac{1}{2} \sin^{-1}\left(\frac{L}{2R}\right)\right\} \tag{12.11}$$

The amount of warp deformation δ_n caused by curing shrinkage is as follows:

$$\delta_n = \sum_{j=1}^{n} \left(\frac{(h'_{1j} + \xi_j)}{h'_{1j}} \cdot \Delta\delta_j + \delta_{j-1} \right) \tag{12.12}$$

However, when $j = 1$, $\delta_{(j-1)} = 0$. The final curing warp δ_{scu} after sufficient time has passed in the case of a laminate made of epoxy resin and steel.

$$\delta_{cu} = k_s \cdot \frac{\xi_n}{h'_{1gel}} \cdot \delta_n \tag{12.13}$$

The curing warpage δ_{pcu} in the case of a laminate consisting of an epoxy resin and a printed circuit board is can be expressed by

$$\delta_{cu} = k_s \cdot \frac{\xi_n}{h'_{1gel}} \cdot \delta_n \tag{12.14}$$

Here, k_s and k_p are the correction coefficients of each laminate, and ξ_n is ξ and h'_{1gel} in the state where the epoxy resin becomes solid after sufficient time has passed ($E = 1.62$ GPa). The correction coefficients k_s and k_p are calculated by comparing them with the experimental values as described later.

12.4.2 Curing Characteristics of Resin Used for Simple Prediction

Figure 12.6 shows the curing behavior of the epoxy resin from liquid to solid in relation to the storage rigidity G', the loss rigidity G'', and the temperature T. A dynamic viscosity evaluation device was used for the measurement.

The sample epoxy resin is loaded between two angled cone plates, and the frequency is 10 Hz, the cone angle is 1.966°, and the cone diameter is 39.98 mm, strain control 1°, the temperature rise rate is 4°C/min.

From Fig. 12.6, it was a viscous fluid at the initial temperature of 25°C, but after that, the viscosity decreased as the temperature rose, and the curing reaction proceeded to change from gelation to a viscoelastic solid. It is considered that this epoxy resin gels at around 88°C (generally referred to as $G' = G''$ temperature) and becomes solid at around 115°C.

12.4.3 Calculation Method Using a Simple Prediction Formula

When calculating the curing warpage using a simple prediction formula, it was assumed that the conversion elastic modulus $E_t(T)$ was used as a value that

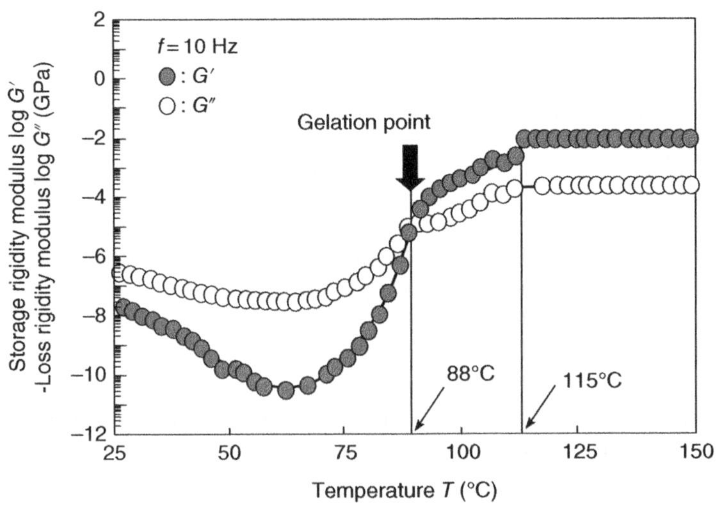

Figure 12.6 Viscoelasticity properties for epoxy resin.

changes with temperature and that it was simply calculated from G' and Poisson's ratio ν by the following formula:

$$G' = \frac{E_r(t)}{2(1+\nu)} \tag{12.15}$$

For the Poisson's ratio ν, a value of 0.3 was used as the solid state. The conversion elastic modulus is an approximation of the temperature change of G' from 88°C to 115°C during the curing process.

Assuming that this rate of change is the same even at temperatures of 115°C or higher, the temperature at which Young's modulus of the solid epoxy resin is 1.62 GPa is obtained, and 128°C is obtained.

Based on this idea, the relationship between the conversion elastic modulus E_t and the temperature T is shown in Fig. 12.7.

Using this calculation method, the curing warpage δ_{cu} is calculated using a simple prediction formula that takes into account the curing behavior of the epoxy resin. In the calculation, the conversion elastic modulus of the epoxy resin is calculated in the range from the gelation start temperature of 88°C to the gelation end temperature of 128°C, and the temperature change of the conversion elastic modulus in this range is calculated by Eq. (12.1). (Curing warpage was predicted as the longitudinal elastic modulus E_1 of the epoxy resin shown in Eqs. (12.15) and (12.9).)

In addition, the coefficient of linear thermal expansion of 0.13% obtained in the preliminary experiment was used as the coefficient of linear expansion α_1 in the

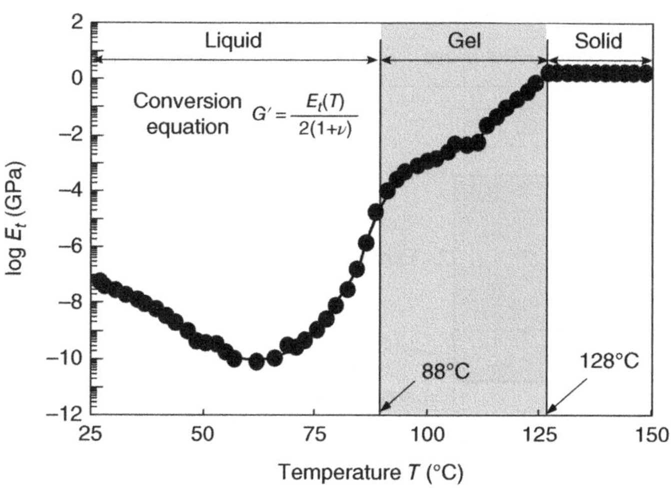

Figure 12.7 Model of Young's modulus.

simple prediction formula. That is, if the shrinkage is 0.13% from 88°C to 128°C, the value of α_1 is 33.4 ppm/K, and this value is used in the calculation of Eq. (12.4).

On the other hand, as the physical property values of Material 2, the values of Young's modulus and linear expansion coefficient of T_g or less shown in Table 12.1 were used.

12.5 Warp Deformation Experiment

12.5.1 Experimental Method

In the experiment, two types of laminates, the epoxy resin and steel as shown in Fig. 12.8, and the epoxy resin and the printed circuit board, were used. The length of the laminate was 150 mm, the width was 15 mm, the thickness of the steel material and the printed circuit board was constant at 0.5 mm, and the epoxy resin thickness was 5, 10, 15, 20, and 25 mm.

Stainless steel was used for the steel material, and FR-4 board with double-sided copper foil, which is often used for electronic parts, was used for the printed circuit board. Then, a liquid epoxy resin was weighed and cast onto a steel material and a printed circuit board to prepare a laminate having the desired thickness.

At this time, the warp deformation state of the laminated body, which changes from moment to moment due to the curing reaction of the epoxy resin, was

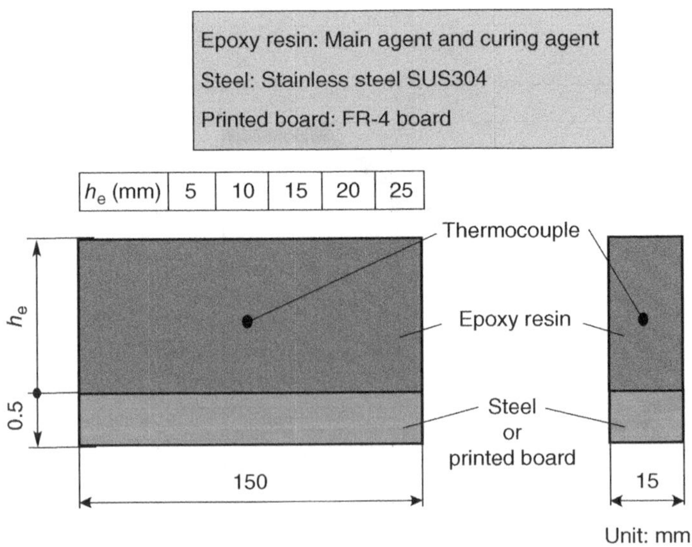

Figure 12.8 Structure and dimensions of the laminate.

sequentially photographed with a CCD camera, and the warp deformation amount was measured from this.

At the same time, the temperature that changes with time due to the thermocouple embedded in the laminate was also measured.

12.5.2 Epoxy Resin Heat Generation Temperature

Figure 12.9 shows the temporal change in the exothermic temperature during the chemical curing reaction of the epoxy resin. From Fig. 12.9, when the epoxy resin thickness is as thin as 5 mm, almost no heat generation is observed, but when it is thick, the heat generation temperature rises over time, and after reaching the maximum temperature, it reaches room temperature of 25°C.

It can also be seen that the thicker the epoxy resin, the shorter the time to reach the maximum temperature and the higher the maximum temperature.

In this way, it is considered that the calorific value due to the curing reaction greatly differs depending on the thickness of the epoxy resin, and when the epoxy resin is thick, the calorific value is large and the curing reaction is rapidly promoted.

12.5.3 Amount of Warp Deformation of the Laminated Body

Figure 12.10 shows the relationship between the amount of warpage deformation and the epoxy resin thickness for a laminate made by combining an epoxy resin and a steel material or an epoxy resin and a printed circuit board.

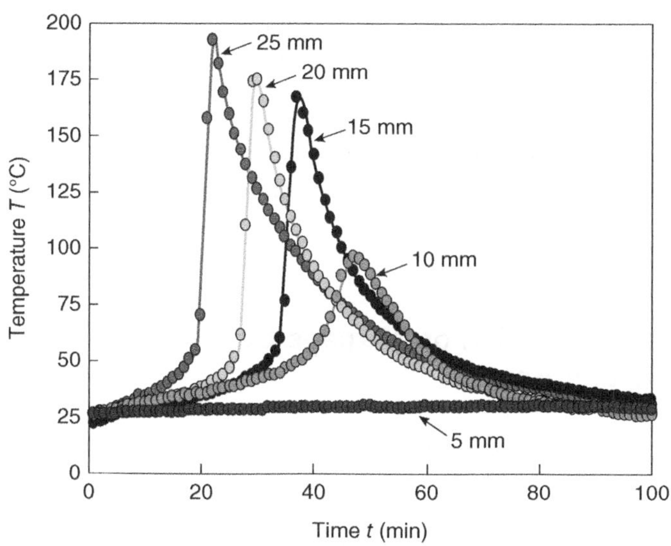

Figure 12.9 Curing temperature by each epoxy resin thickness.

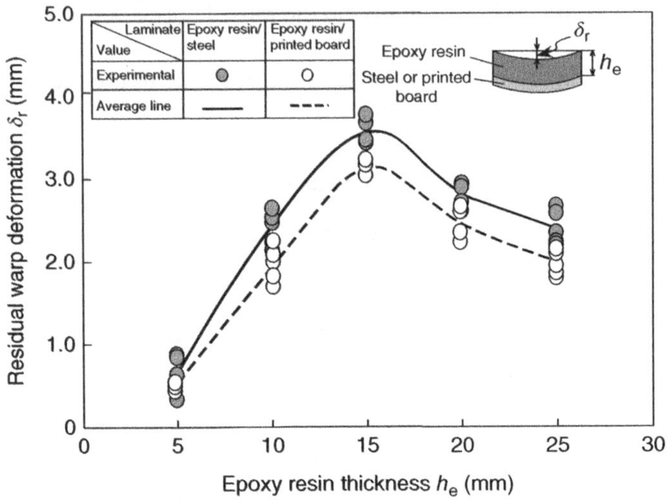

Figure 12.10 Experimental warp deformation of each laminated body.

From Fig. 12.10, the amount of warpage deformation differs greatly depending on the epoxy resin thickness in both laminates, and within the range of this experiment, the residual warp deformation amount is maximum when the epoxy resin thickness is 15 mm, and the combination of epoxy resin and steel material.

In this way, the reason why the amount of warp deformation differs depending on the difference in the constituent materials of the laminate and the thickness of the epoxy resin is due to the thermal and mechanical properties such as the elastic modulus, linear expansion coefficient, and glass transition temperature of the constituent materials.

It is an interesting phenomenon that the magnitude of the resulting thermal expansion/contraction force seems to be mutually and closely related to the bending rigidity of the entire laminate.

12.6 Theoretical Prediction of Warpage Deformation due to Hardening and Heat

The aforementioned combinations of laminated bodies were examined, and a series of results regarding the amount of warp deformation of these combinations are summarized in Table 12.2.

From Table 12.2, the total amount of warp deformation of the laminate caused by both the curing reaction (curing warp: δ_{cu}) and the heat load (heat warp: δ_{th})

Table 12.2 Comparison between theoretical and experiment values.

Unit: mm, ( ):%

Laminate / Item	Epoxy resin/steel											
	$b = 5$							$b = 15$				
h	5	10	15	20	25	30	35	5	10	15	20	25
δ_{cu}	0.160 (62)	0.138 (49)	0.140 (44)	0.143 (38)	0.143 (40)	0.141 (41)	0.137 (40)	0.497 (72)	0.427 (18)	0.435 (12)	0.443 (16)	0.442 (18)
δ_{th}	0.097 (38)	0.146 (51)	0.178 (56)	0.236 (62)	0.211 (60)	0.204 (59)	0.205 (60)	0.191 (28)	2.004 (82)	3.067 (88)	2.346 (84)	1.972 (82)
δ_{tc}	0.257 (100)	0.284 (100)	0.318 (100)	0.379 (100)	0.354 (100)	0.345 (100)	0.342 (100)	0.688 (100)	2.431 (100)	3.502 (100)	2.789 (100)	2.414 (100)
δ_{ex}	0.238	0.312	0.369	0.374	0.335	0.312	0.311	0.775	2.466	3.382	2.796	2.327
Error[a]	0.08	0.09	0.14	0.01	0.06	0.11	0.10	0.11	0.01	0.04	0.01	0.04

a $\frac{\delta_{ex} - \delta_{tc}}{\delta_{ex}}$.

should be in good agreement with the experimental value with an error of about 1–5%. Became clear.

In addition, from a series of theoretical and experimental studies, although within the scope of this study, the proportion of the total warp deformation of the laminate due to the curing reaction is about 15–20%, and most of it was presumed to be warpage deformation due to heat load.

Viscoelastic analysis with VESAP and experiments are conducted to identify the amount of warpage deformation that occurs during the curing process of epoxy resin in a laminate consisting of epoxy resin and steel, and epoxy resin and printed circuit board.

We examined both sides. As a result, it was clarified that the ratio and contribution rate of the curing reaction to the warp deformation of the laminated body is estimated to be about 15–20%.

This study was conducted with a general-purpose two-component epoxy resin, and it is expected that it will be different from the resin used for actual semiconductor packages, electronic components, and board mounting.

In addition, the warp deformation was predicted by introducing the concept of conversion elastic modulus based on the experimental data of the storage rigidity and the loss rigidity when the epoxy resin undergoes a phase change from a liquid to a solid. We think that this idea can be generally applied, but it is necessary to examine the effectiveness of this prediction formula for resins used in various actual electronic devices.

13

Changes in Material Properties and Deformation Behavior Due to Thermal Degradation

13.1 Purpose and Background

Many electronic components are used in electronic devices, and there are a wide variety of large-scale integration (LSI) packages among them. In recent years, CSP-BGA-type packages have become the mainstream for this LSI package due to the demand for thinness and small size, and the development competitiveness of each company is intensifying. The main aim of development is high reliability, high functionality, and reduction of manufacturing cost.

On the other hand, the CSP-BGA package has an extremely thin structure, and is subject to various heat loads during manufacturing, such as sealing the LSI chip with resin and loading a soft resin such as elastomer between the LSI chip and the substrate.

It is required to optimize the resin and structure to prevent problems such as warpage deformation, interface peeling and cracking, and the thermal process during manufacturing.

By the way, it is no exaggeration to say that the most important path in optimized design is the characteristics of the resin used. Therefore, it is indispensable to understand the basic contents of how the mechanical properties of the resin change, especially when the thermal load is continuously applied.

From such a background, I grasped by basic experiments how the thermal and mechanical so-called viscoelastic properties of thin tape-shaped or film-shaped resin change due to long-term heat load.

Specifically, two experimental cases will be described as follows.

Figure 13.1 shows the necessity of development and the outline of the background.

Polymeric Materials for Electronic Packaging, First Edition. Shozo Nakamura.

Published 2023 by John Wiley & Sons, Inc.

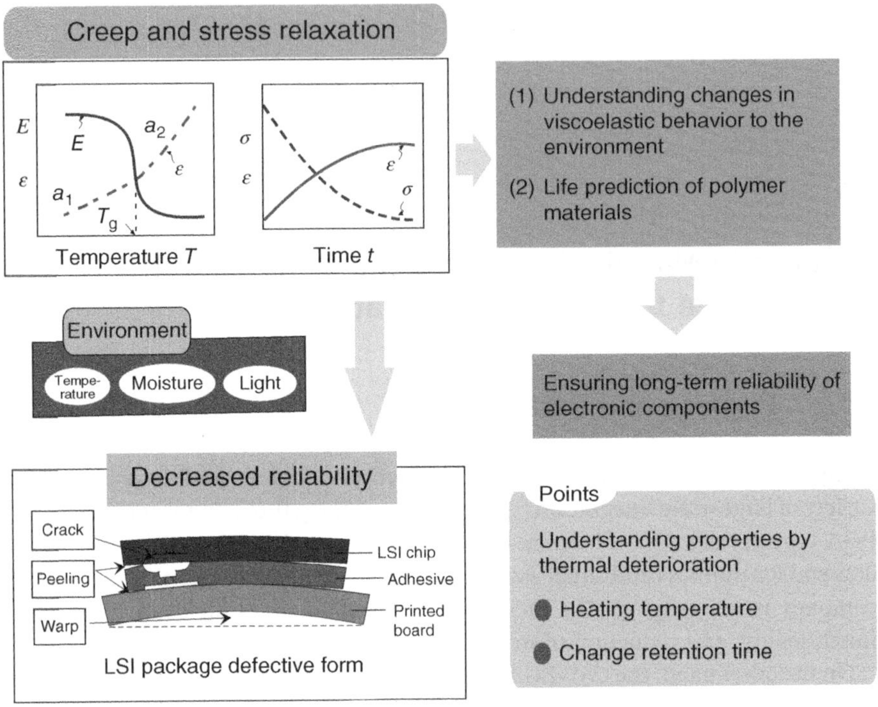

Figure 13.1 Necessity and background of development for thermal deterioration.

13.2 Experimental Case I

13.2.1 Specimen and Heat Load

Figure 13.2 shows the specimen used and the heat load in the experiment.

13.2.2 Method of Experimental (I)

Physical property changes of the materials after thermal aging shown in Fig. 13.2 were investigated by dynamic viscoelasticity test and normal sexual tensile test. The dynamic viscoelastic properties were investigated from the dynamic viscoelasticity test as described in Chapter 6.

That is, the test piece was fixed between chucks, and the temperature was raised from 20°C to 200°C with a constant frequency of 10 Hz.

Then, measured ① storage modulus E', ② loss modulus E'', and ③ loss tangent $\tan\delta$. The results are shown in Figs. 13.3–13.5.

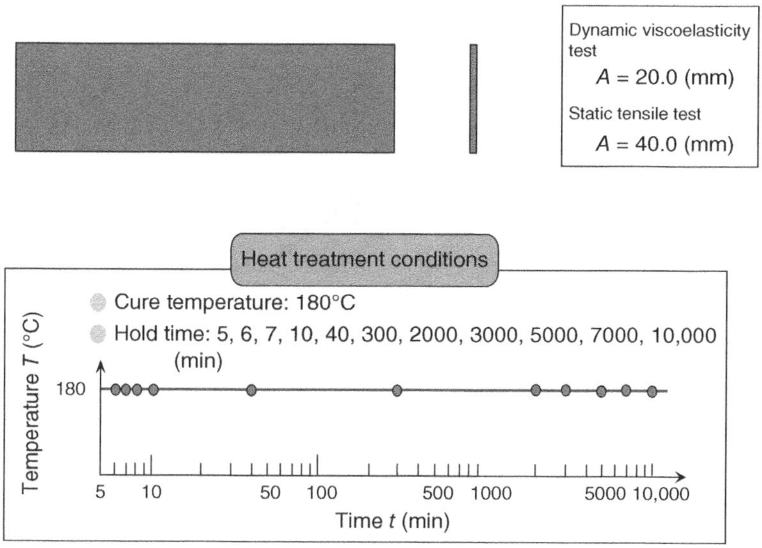

Figure 13.2 Specimen and heat load conditions.

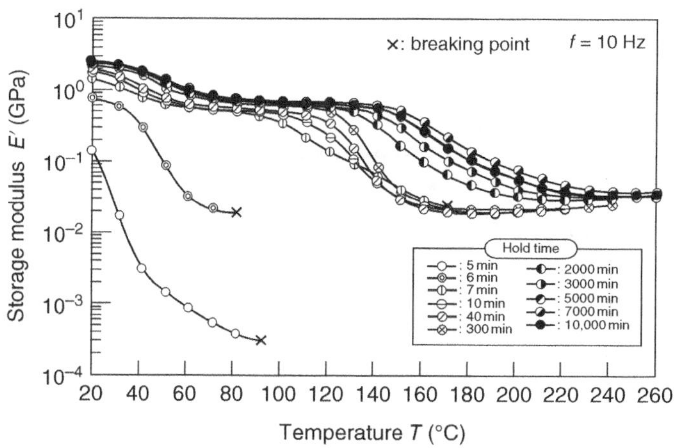

Figure 13.3 Storage modulus E' vs. temperature.

On the other hand, the static tensile test was conducted at a temperature of 20°C and a tensile speed of 5 mm/min. As a result, ① stress–strain curve, ② tensile strength, and ③ elastic modulus were measured and shown in Figs. 13.7–13.9.

Also, a photograph of the fracture surface of the test piece after thermal aging is also shown.

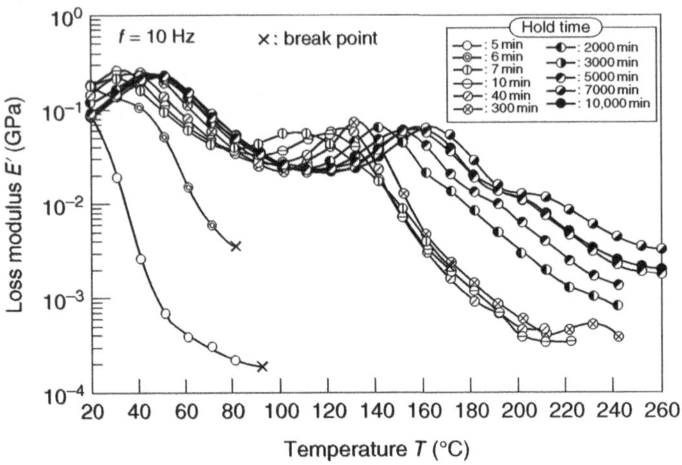

Figure 13.4 Loss modulus E'' vs. temperature.

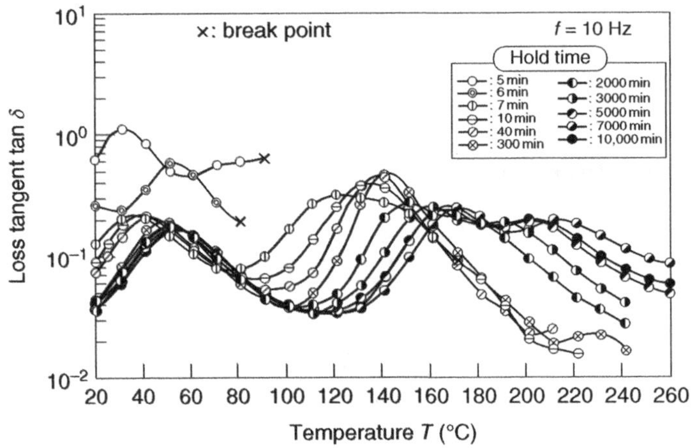

Figure 13.5 Loss tangent vs. temperature.

13.2.3 Results of Experimental (I)

i) Dynamic viscoelastic properties
   ① Storage modulus E' (Fig. 13.3)
   ② Loss modulus E'' (Fig. 13.4)
   ③ Loss tangent (tan δ) (Fig. 13.5)
   ④ Glass transition temperature (Fig. 13.6)

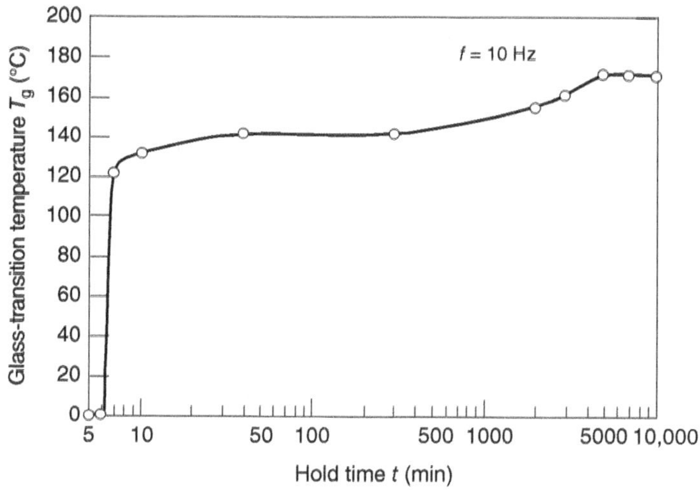

Figure 13.6 Glass transition temperature vs. holding time.

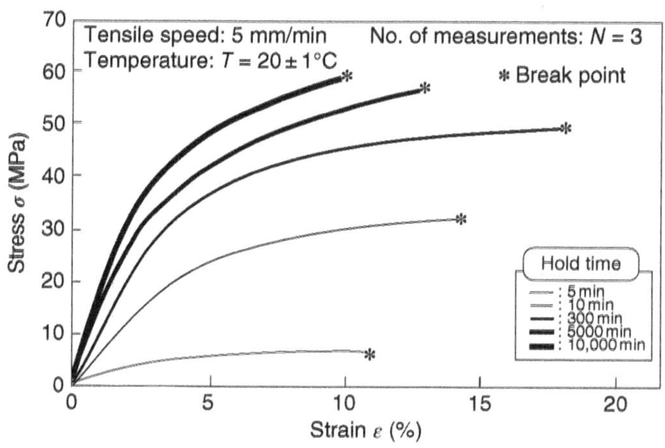

Figure 13.7 Stess–strain curve vs. holding time.

ii) Static tensile properties

Figures 13.7–13.9 are the results of static tensile properties, showing stress–strain curves, tensile strength, and longitudinal modulus.

① Stress–strain characteristics (Fig. 13.7)

② Tensile strength (Fig. 13.8)

③ Elastic modulus (Fig. 13.9)

iii) Fracture surface after thermal aging

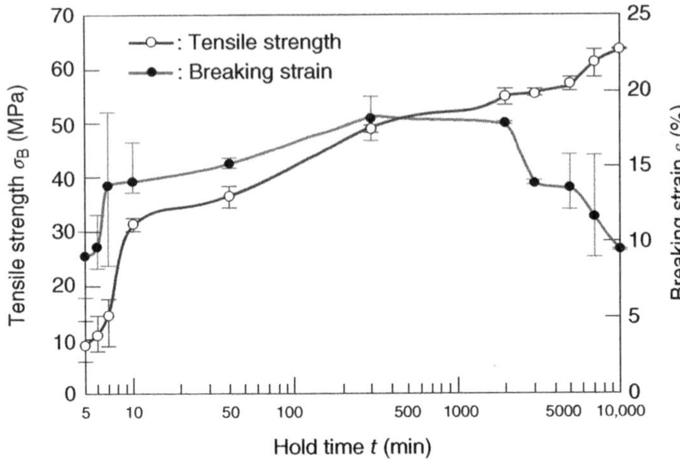

Figure 13.8 Strength of specimen vs. holding time.

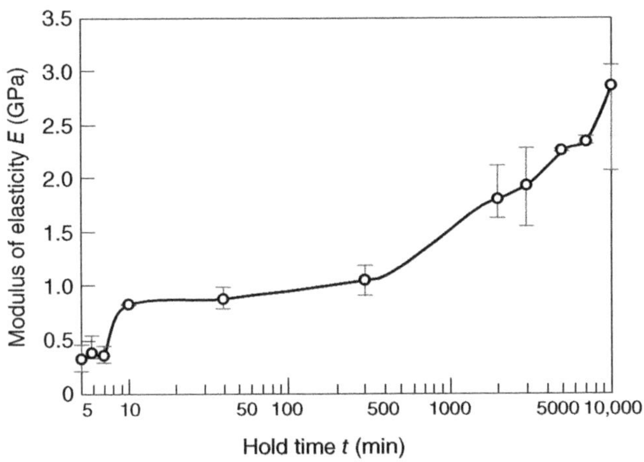

Figure 13.9 Elastic modulus vs. holding time.

The following photographs show the fracture surface for each holding time. From Fig. 13.10, it can be seen that the longer the holding time, the smoother the cross-section, indicating that the material is embrittled.

Note that the numbers at the top of the photograph are the holding times under heat load.

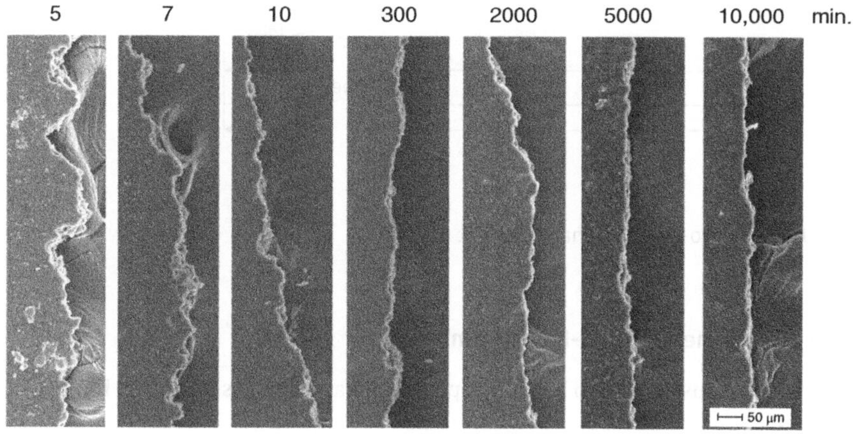

Figure 13.10 Fracture surface after thermal aging.

13.2.4 Conclusion of Experiment in Case of (I)

1) As the holding time increases,
   ① The value of the storage elastic modulus becomes high.
   ② The shape of the extreme value of the loss modulus changes from broad to sharp, and finally broad.
   ③ The extreme value of the loss tangent changes from increasing to decreasing after the holding time of 300 minutes.
2) The fracture stress and longitudinal elastic modulus increase as the holding time increases, but the fracture strain starts to increase and then decrease after the holding time of 300 minutes.
3) As the holding time increases, the fracture shape becomes a smooth fracture surface. Experimental dynamic viscoelastic properties and tensile properties in the curing process of adhesives for electronic components.

We examined and clarified the following matters.

13.3 Experiment of Case II

In this experiment, the changes in viscoelastic properties of the two-layer laminate were investigated when the thermal load was changed.

The results are described here.

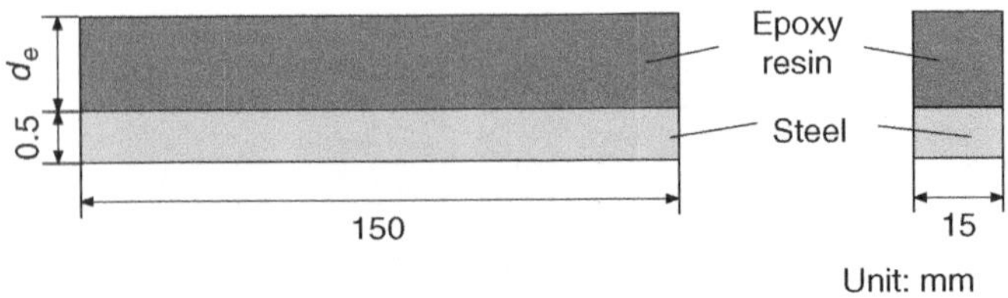

Figure 13.11 Two-layer laminate for heat load.

13.3.1 Specimen of Two-Layer Laminate

The laminate consists of two layers of epoxy resin and steel as shown in Fig. 13.11.

13.3.2 Properties of Specimen

Heat loads of 0, 30, 45, and 60 minutes were applied to the laminate, and their physical properties are shown in Table 13.1.

13.3.3 Results of Experimental in Case of (II)

The following Figs. 13.12–13.15 shows the change in viscoelastic properties when the above thermal load is applied.

i) Storage modulus E' (Fig. 13.12)
ii) Loss tangent (Fig. 13.13)
iii) Relaxation modulus E_r (Fig. 13.14)
iv) Time–temperature shift factor $a_{T0}(T)$ (Fig. 13.15)
v) Result of Warp deformation of laminate

Figures 13.16–13.20 show a series of warpage deformation behaviors when the two-layer laminate with the above thermal load is heated from room temperature to 130°C and then cooled again to room temperature.

13.3.4 Conclusion of Experiment in Case of (II)

As described earlier, the changes in the viscoelastic properties of the resin, when a constant heat load of 260°C was applied to the resin constituting the laminate and the holding times, were changed by four types were investigated.

Next, a series of warp deformation behaviors from heating to cooling were experimentally clarified for the two-layer laminate using these resins.

Table 13.1 Properties of laminate.

Property / Material	Thermal conductivity λ (W/m K)	Specific heat C (kJ/kg K)	Density ρ (kg/m^3)	Glass-transition temperature T_g (°C)	Expansion coefficient α (10^{-6}/K)		Modulus E (GPa)	
					$T \leqq T_g$	$T_g < T$	$T \leqq T_g$	$T_g < T$
Epoxy resin (before heat load)	0.186	1.423	1220	105	81.0	168.4	1.96	0.0305
(260°C 30 min)				100	73.0	104.9	1.61	0.0263
(260°C 45 min)				85	75.3	104.7	1.67	0.0567
(260°C 60 min)				82	73.2	75.9	1.61	0.0261
Steel	14.7	0.452	7870	—	11.7		197	

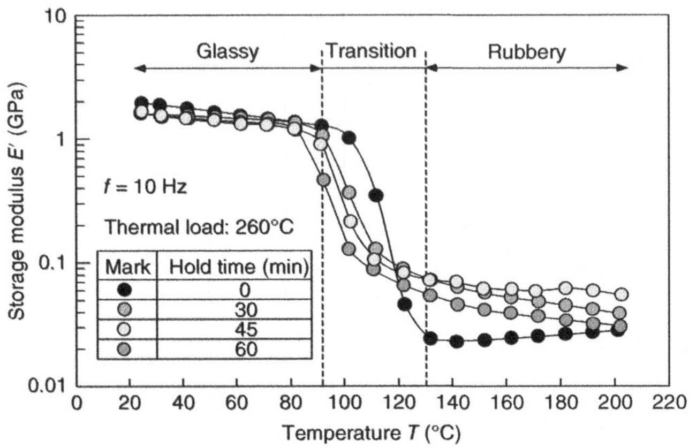

Figure 13.12 Storage modulus of epoxy resin.

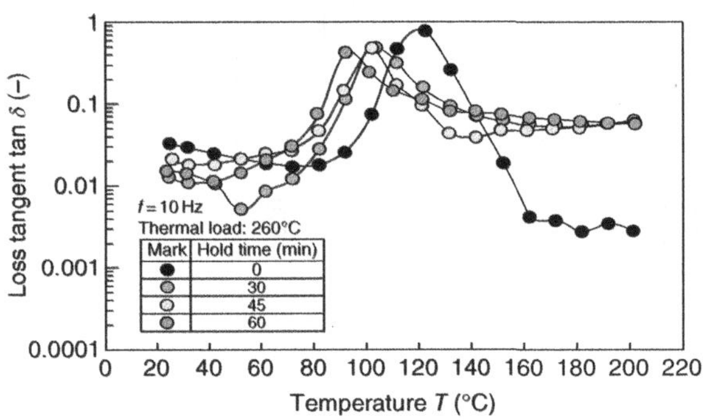

Figure 13.13 Loss tangent of epoxy resin.

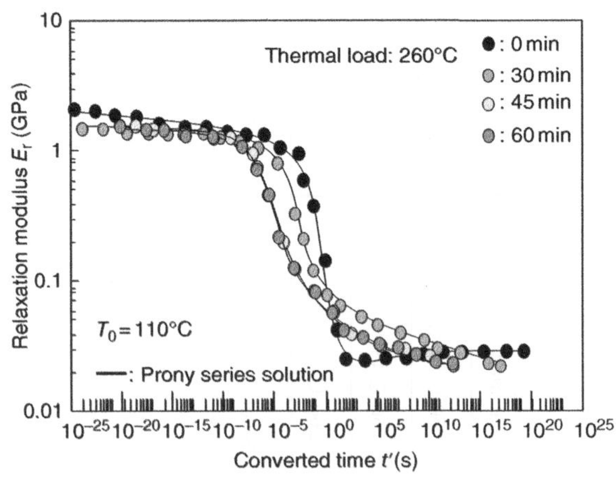

Figure 13.14 Master curve of relaxation modulus.

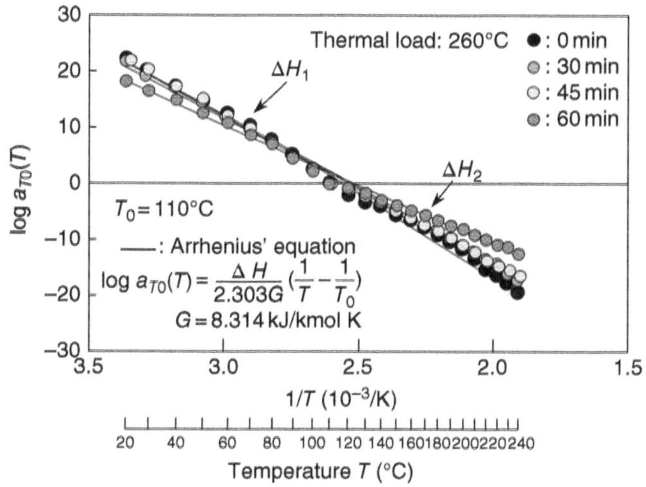

Figure 13.15 Shift factor of epoxy resin.

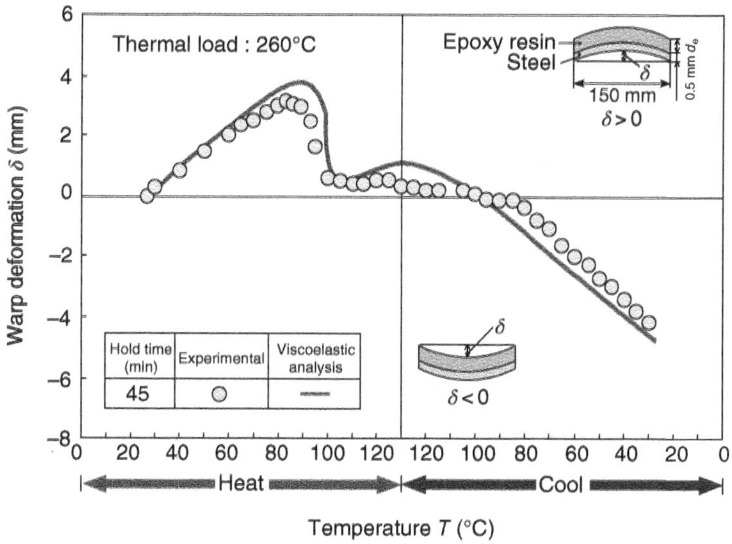

Figure 13.16 Warp deformation behavior of laminate after thermal load (hold time: 0 minute).

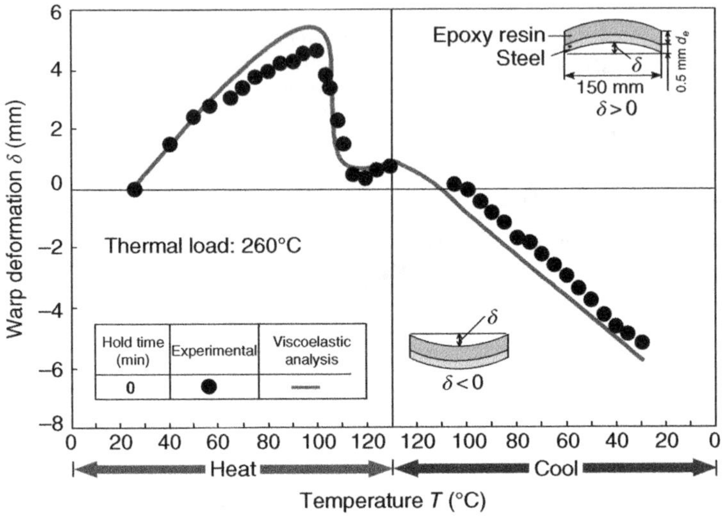

Figure 13.17 Warp deformation behavior of laminate after thermal load (hold time: 30 minutes).

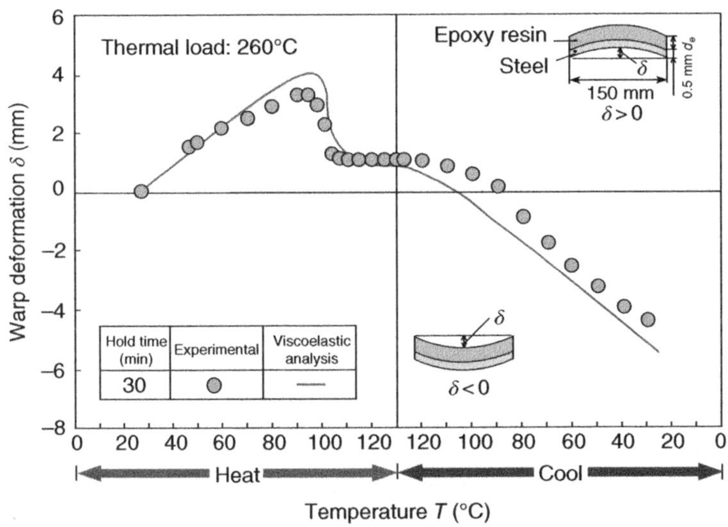

Figure 13.18 Warp deformation behavior of laminate after thermal load (hold time: 45 minutes).

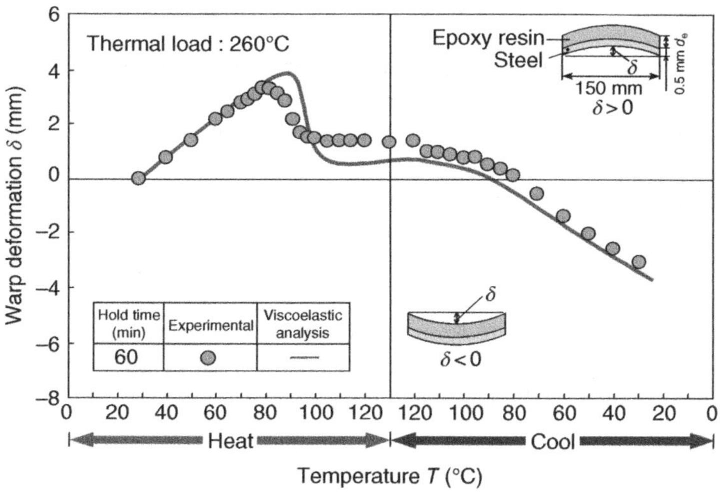

Figure 13.19 Warp deformation behavior of laminate after thermal load (hold time: 60 minutes).

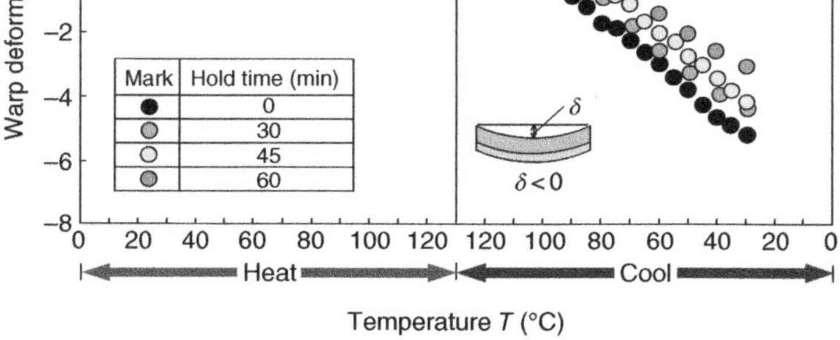

Figure 13.20 Warp deformation behavior of laminate after a combination of various holding times.

Details are shown in each figure, but the following was clarified.

1) As the holding time becomes longer, the viscosity of the resin decreases, showing a tendency of so-called brittleness.
2) The longer the holding time, the higher the glass transition temperature of the resin.
3) Regarding the warp deformation of the laminated body, the resin side increases in a convex state during heating and decreases with cooling, and conversely, the resin side becomes a concave state at room temperature, resulting in a reverse warp phenomenon.
4) As the resin deteriorates, the rate of change in the amount of warpage deformation during heating and cooling tends to decrease. That is, the amplitude of the warp deformation of the laminated body is small.

14

Simple Evaluation Method for Deformation of Viscoelastic Body

14.1 Introduction

In this chapter, we examined a method for easily predicting and evaluating the amount of warpage deformation when a heat load is applied to a laminate containing resin.

Specifically, for epoxy resin/printed circuit boards, which are simple models of electronic devices such as semiconductor packages, and epoxy resin/steel laminates, the amount of warpage deformation is easily predicted at the expense of some rigor. A simple formula was examined.

Then, the amount of warpage deformation (hereinafter referred to as the simplified value) obtained from this simplified formula is compared with the experimental value and the value obtained from the rigorous Visco-Elastic Stress Analysis Program (VESAP) analysis (hereinafter referred to as the theoretical value), and the simplified formula of the present proposal is used.

Describe the results of examining the usefulness. This simplified formula has application limits and cannot be applied to all products, but this concept may be applicable to electronic products containing resins. It is a reference for corporate design engineers when they want to prioritize quick solution directions and design guidelines over quantitative accuracy and accuracy in some cases.

14.2 Derivation of Simple Formula

14.2.1 Concept of Derivation

Generally, in a series of processes until a liquid epoxy resin becomes a solid, two phenomena of curing shrinkage and heat shrinkage occur. Regarding this type of problem, the author wrote.

Polymeric Materials for Electronic Packaging, First Edition. Shozo Nakamura.

Published 2023 by John Wiley & Sons, Inc.

The ratio of warp deformation due to curing shrinkage to the total warp deformation amount (warp due to heat shrinkage + warp due to curing shrinkage) varies depending on the resin thickness, but is approximately 15–20%. And most of it is the amount of warp deformation caused by heat. It was reported in Chapter 12.

From the above, it is considered that the amount of warp deformation can be obtained very easily only by the amount of warp deformation due to heat shrinkage of the resin show.

In deriving the simplified formula, consider a laminate in which the length L and width b are constant and the thicknesses of material 1 and material 2 are h_1 and h_2, respectively, as shown in Fig. 14.1.

Figure 14.2 shows the storage elastic modulus E (upper figure) and the linear expansion coefficient α (lower figure) among the viscoelastic properties of epoxy resins.

In the simple formula, as shown by the solid line in Fig. 14.2, the physical property values of the viscoelastic material that changes intricately with this temperature are shown by two straight lines (less than T_g: glass-like region and T_g or more: rubber) with the glass transition temperature T_g as the boundary. Approximately in the shape region.

Next, in order to derive the simple formula, we tried two ideas, such as displaying it as a simple formula using the basic formula based on the idea of conventional strength of materials. Therefore, it is possible that there may be formula configurations or theoretical jumps that do not necessarily hold strictly.

In addition, from the preliminary experiment, the epoxy resin phase-changed from liquid to solid by a chemical reaction, and the changed state gradually changed to a solid state from the center to both ends of the laminate, and from this point on, the laminate became a solid.

It was found that warpage deformation occurred. Therefore, in the simple derivation, I thought that the warpage deformation occurs from the time when both ends of the laminated body are solidified, and the warp deformation is caused by the difference in the storage elastic modulus E and the linear expansion coefficient α of both materials, and the flexural rigidity of the laminated body.

I thought it was based on. Furthermore, the starting temperature of warpage deformation was set to the curing temperature T_c at which the chemical

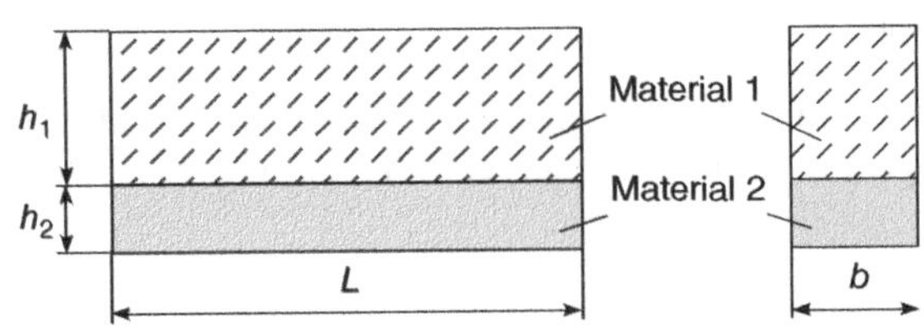

Figure 14.1 Laminated shape and dimensions.

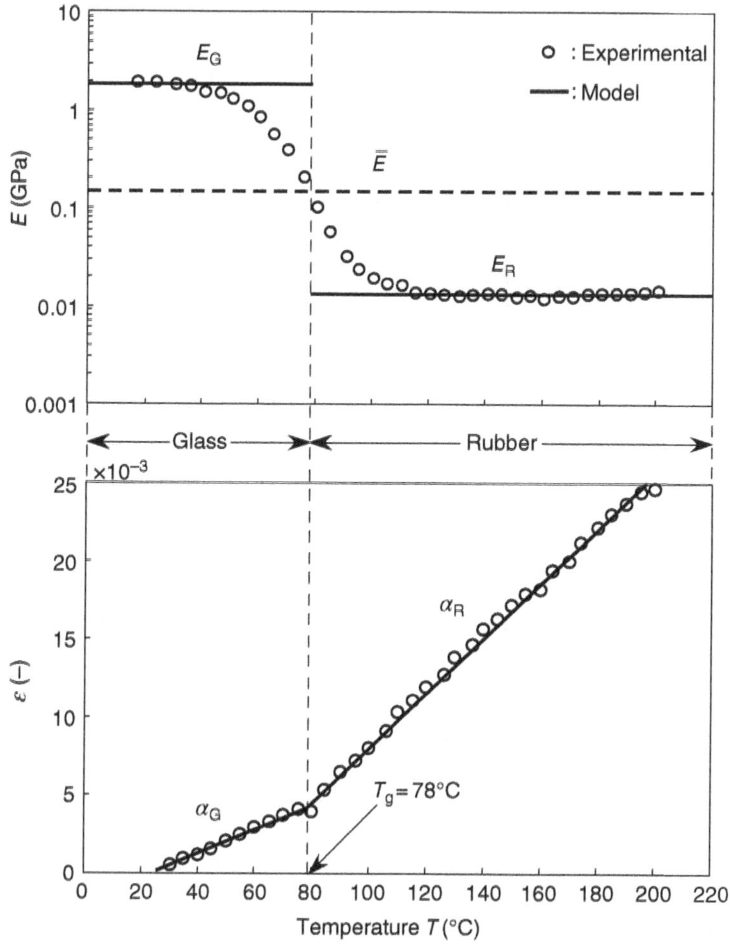

Figure 14.2 Epoxy resin viscoelastic properties and modeling (upper: modulus E, lower: expansion coefficient α).

reaction of the epoxy resin was almost completed. The curing temperature T_c will be described later.

14.2.2 Formulation of Simple Formula

When the temperature change from the curing temperature T_c to the room temperature T_e is applied to the laminate, the formula for calculating the thermal stress σ generated due to the thermal shrinkage of each of the material 1 and the material 2 is calculated with the glass transition temperature T_g as the boundary. Expressed in two formulas.

The derivation process is shown below. Here, the storage elastic modulus E and the coefficient of linear expansion α were defined as E_G and α_G for those below the glass transition temperature (glass-like region) and E_R and α_R for those above the glass transition temperature (rubber-like region), respectively.

i) When $T_e < T_c < T_g$
Since the thermal stress σ is the storage elastic modulus E_G and the linear expansion coefficient α_G, it can be expressed by the following equation.

$$\overline{E_i} = E_{Gi}, \quad i = 1, 2 \tag{14.1}$$

$$\sigma_i = (T_c - T_e) \cdot \alpha_{Gi} \cdot E_{Gi}, \quad i = 1, 2 \tag{14.2}$$

ii) When $T_e < T_g \leq T_c$
The effect of the temperature difference between $(T_c - T_g)$ depending on the epoxy resin thickness on the storage elastic modulus E is expressed as an average value. In this way, we attempted to easily express the thermoviscoelastic property in which the storage elastic modulus E changes significantly with time and temperature.
Here, as shown by the broken line in Fig. 14.2, the storage elastic modulus E is the value of the average storage modulus E in the temperature interval from the curing temperature T_c to the room temperature T_e.

Therefore, the thermal stress σ in each temperature region can be expressed by the following equation.

$$\overline{E_i} = \frac{(T_c - T_g) \cdot E_{Ri} + (T_g - T_e) \cdot E_{Gi}}{T_c - T_e}, \quad i = 1, 2 \tag{14.3}$$

$$\sigma_i = (T_c - T_g) \cdot \alpha_{Ri} \cdot E_{Ri} + (T_g - T_e) \cdot \alpha_{Gi} \cdot E_{Gi}, \quad i = 1, 2 \tag{14.4}$$

If the bending moments of materials 1 and 2 generated by the thermal stresses of Eq. (14.2) and (14.4) are M_1 and M_2, respectively, these bending moments are the thermal stress σ and the section modulus of material 1 and material 2. It can be expressed by the following equation from Z_1 and Z_2.

$$M_i = \sigma_i \cdot Z_i, \quad i = 1, 2 \tag{14.5}$$

$$Z_i = \frac{b \cdot h_i^2}{6}, \quad i = 1, 2 \tag{14.6}$$

Next, Eqs. (14.5) and (14.6) are equations used for a single beam, but in the case of a laminate in which two materials are adhesively bonded to each other, it is considered that they are constrained by each other's materials.

Therefore, it was considered that the thermal warp deformation of the laminate was caused by the difference between the bending moments M_1 and M_2. Based on this, the bending moment involved in thermal warpage deformation was defined as M_d, which was expressed by the following equation.

$$M_d = M_1 - M_2, \quad i = 1, 2 \tag{14.7}$$

Also, if the geometrical moment of inertia of material 1 and material 2 is I_1 and I_2, respectively,

$$I_i = \frac{b \cdot h_i^3}{12}, \quad i = 1, 2 \tag{14.8}$$

The curvature κ generated in the laminate can be expressed by the following equation from the conventional bending theory.

$$\kappa = \frac{M_d}{\left(\overline{E_1} \cdot I_1 + \overline{E_2} \cdot I_2\right)} \tag{14.9}$$

Finally, the residual thermal warp deformation amount δ of the laminated body can be calculated by the following equation.

$$\delta = \frac{L}{2} \tan\left\{\frac{1}{2} \sin^{-1}\left(\frac{L\kappa}{2}\right)\right\} \tag{14.10}$$

14.3 Practical Method

In the experiment, the epoxy resin/printed circuit board shown in Fig. 14.1 and the laminate made of epoxy resin/steel were used.

The constituent materials of the laminate are two types of epoxy resins with different properties, in which the main agent and the curing agent are mixed in the desired ratio in the material 1, and the printed circuit board (FR-4 substrate composed of glass fiber and epoxy resin) is used in the material 2, Steel material was used.

Here, the shape and dimensions of the laminate were constant, with a length L of 150 mm and a width of 15 mm, and the thickness h_2 of the printed circuit board was 1.0 and 1.6 mm. There were a total of 15 types of epoxy resin thickness h_1 changed from 5 to 25 mm at 5 mm intervals, compared to three types with a thickness h_2 of 0.5 mm.

The laminate was prepared by injecting a predetermined amount of liquid epoxy resin onto the printed circuit board and steel material. The warp deformation state during this manufacturing process was sequentially photographed with a CCD camera, and the amount of warp deformation was measured using measurement software.

The curing start temperature T_c of the laminate was measured from the temperature that changes from moment to moment by loading a thermocouple in the

Table 14.1 Viscoelastic properties of materials.

Property / Material	Thermal conductivity λ (W/m K)	Specific heat C (kJ/kg K)	Density ρ (kg/m^3)	Glass transition temperature T_g (°C)	Expansion coefficient α (10^{-8}/K)		Modulus E (GPa)	
					$T \leq T_g$	$T_g < T$	$T \leq T_g$	$T_g < T$
Epoxy resin A	0.186	1.423	1220	108	80.6	161.6	1.53 ~ 0.015	
Epoxy resin B				78	72.0	176.6	2.05 ~ 0.015	
FR-4 (1.0 mm)	0.32	1.256	1810	147	13.4	10.9	10.7 ~ 5.29	
FR-4 (1.6 mm)				137	9.2	2.0	7.11 ~ 2.92	
Metal	25.11	0.460	7780	—	11.7		204.0	

center of the epoxy resin. At the same time, a video camera was used to photograph the striped pattern appearing on the laminate during the curing process using a photoelastic device.

In addition, the dynamic viscoelasticity measuring device Rheogel E-4000 and the thermomechanical analyzer TMA-50 were used to measure the viscoelastic properties of various materials.

Table 14.1 shows the measured viscoelastic properties and thermal properties. These values were used for the viscoelasticity analysis by VESAP described above. Since the basic formulas for viscoelasticity analysis have been described above, they will be omitted.

14.4 Determining the Curing Temperature of the Resin

Figure 14.3 shows the measurement results of the temperature change in temperature at the curing temperature of the epoxy resin.

Here, the change in the stress stripe pattern observed by the photoelastic device is the temperature at which the stripe pattern begins to appear in the center of the epoxy resin at T_t marked with ○ in Fig. 14.3, and the stripe pattern appears over the entire epoxy resin at T_c marked with □. Indicates the temperature that has appeared.

From the observation results of this photoelastic device, it was found that the epoxy resin was still in a gel state and could not be transformed into a solid at the maximum exothermic temperature T_{max}. Then, it was clarified that the curing

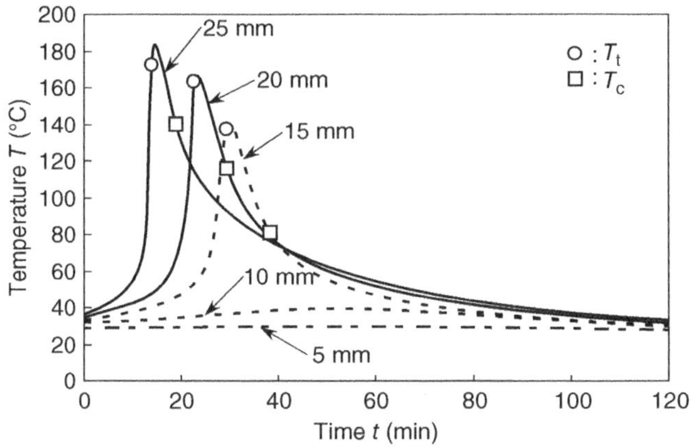

Figure 14.3 Relationship between epoxy resin thickness and heat generation temperature.

temperature T_c marked with □ is lower than the maximum exothermic temperature T_{max}.

Here, the authors consider that the time when the striped pattern seen inside the epoxy resin appears in the entire laminate is the completion of the curing shrinkage, which is the change from liquid to solid, and the accompanying curing warpage. No thermal warpage deformation occurs in the laminate.

14.5 Effect of Epoxy Resin Thickness on Heat Generation Temperature

I thought, that is, the curing temperature T_c is the temperature at which the entire laminate is solidified. This temperature was used in a simple formula. In addition, from Fig. 14.3, it can be seen that the time required from the maximum heat generation temperature T_{max} to the curing temperature T_c is shorter as the epoxy resin becomes thicker. It is considered that when the epoxy resin is thick, the curing reaction is rapidly accelerated, the calorific value is increased, and the time to reach the curing temperature T_c is shortened.

Figures 14.4 and 14.5 show the experimental and theoretical values of the amount of warpage deformation of various laminates with different epoxy resin thicknesses, as well as simple values. Here, the plot is an experimental value, the broken line is a theoretical value, and the solid line is a simple value. Figure 14.4 shows a simple value calculated by using the maximum heat generation temperature T_{max} of the epoxy resin in a simple manner, comparing it with the experimental value and the theoretical value.

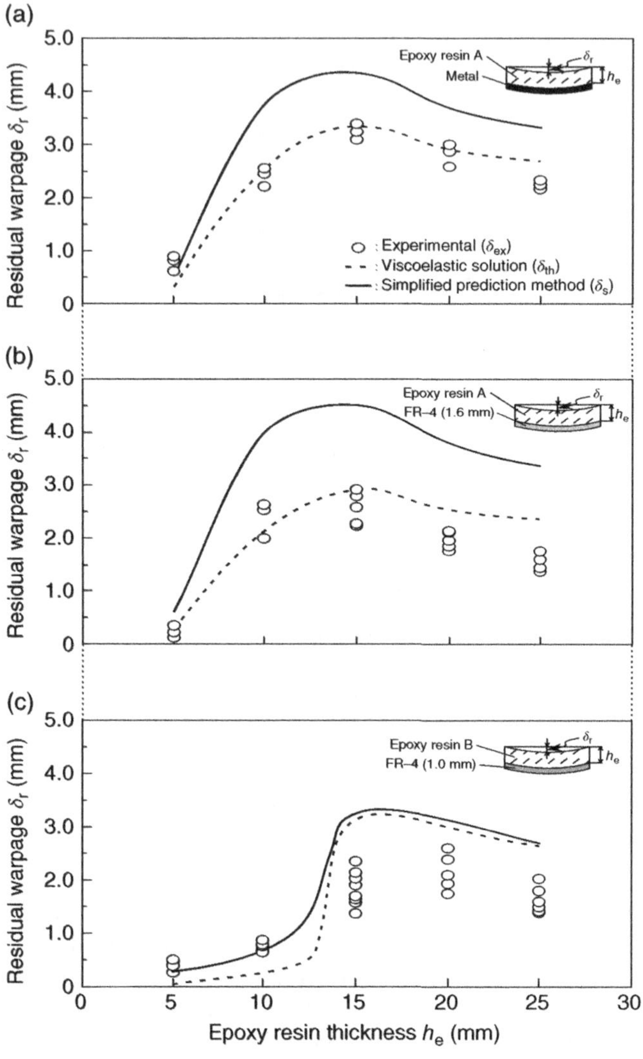

Figure 14.4 Residual warp deformation amount (in the case of the maximum temperature). (a) Epoxy resin A/SUS 420J2 (0.5 mm), (b) epoxy resin A/FR-4 (1.6 mm), and (c) epoxy resin B/FR-4 (1.0 mm).

From Fig. 14.4, in the case of (a) Epoxy resin A/Steel and (b) Epoxy resin A/FR-4 (1.6 mm), the simple values of the amount of warpage deformation with respect to the increase in epoxy resin thickness are experimental values and theories. The absolute value is different from the value, but the tendency is the same. In addition,

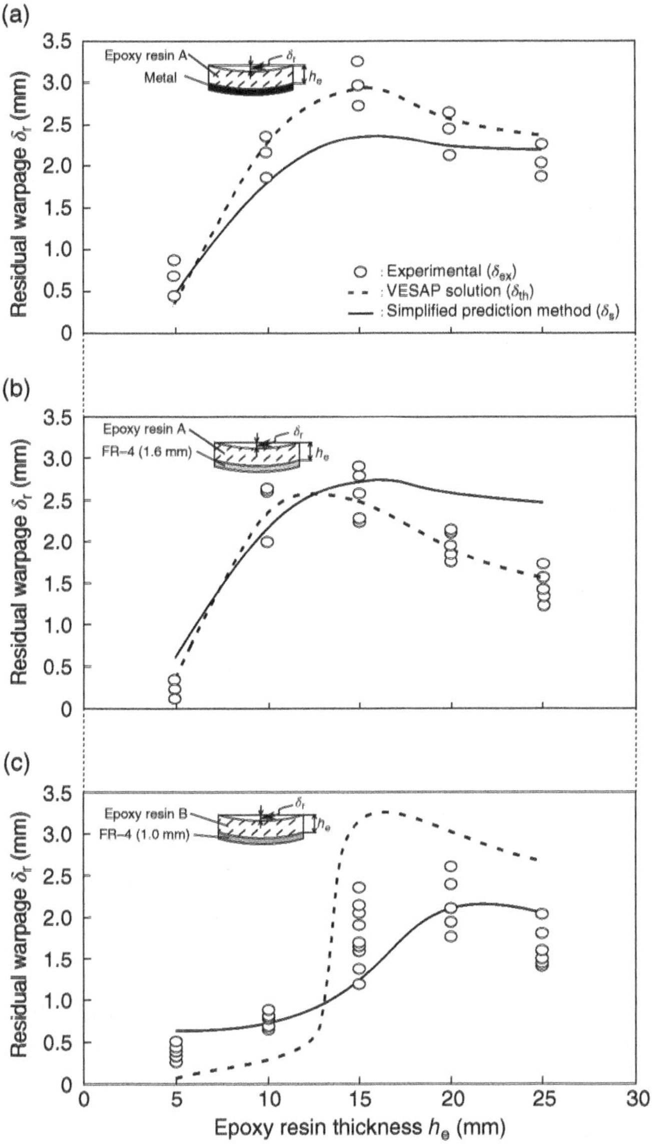

Figure 14.5 Residual warp deformation amount (in the case of curing temperature). (a) Epoxy resin A/metal (0.5 mm), (b) epoxy resin A/FR-4 (1.6 mm), and (c) epoxy resin B/FR-4 (1.0 mm).

in the case of (c) Epoxy resin B/FR-4 (1.0 mm), it can be seen that the simplified values are almost the same as the theoretical values.

On the other hand, Fig. 14.5 shows a simple value calculated by using the curing temperature T_c of the epoxy resin in a simple manner, comparing it with the experimental value and the theoretical value.

From Fig. 14.5, it can be seen that the simple values generally match the absolute values and tendencies of the warp deformation amount with the experimental values and theoretical values.

That is, by applying the curing temperature T_c of the epoxy resin to the simple formula, the simple value of the warp deformation amount matches the experimental value, and the simple value obtains the warp deformation amount accurately compared to the result of Fig. 14.4 above.

I found that I could do it. Therefore, it can be said that it is desirable to use the curing temperature T_c rather than the maximum exothermic temperature T_{max} in the simple formula for determining the amount of warpage deformation of the laminate made of epoxy resin that has become a solid state through the curing process.

From the above, the simple formula proposed here approximates the physical property values such as the storage elastic modulus and the linear expansion coefficient with two straight lines with the glass transition temperature of the viscoelastic material as the boundary, and the temperature at which the warp deformation starts. It was clarified that the amount of warpage deformation can be easily predicted by using the curing temperature T_c of the epoxy resin.

14.6 Conclusion

A simple formula was proposed to easily predict and evaluate the amount of warpage deformation of an epoxy resin and a printed circuit board that exhibit viscoelastic behavior, and a two-layer viscoelastic laminate composed of an epoxy resin and a steel material. The simple values obtained from this simple formula were compared and examined using the experimental values and the theoretical values obtained by analysis based on the viscoelasticity theory, and the following conclusions were obtained.

1) An evaluation method for determining the amount of warp deformation of a viscoelastic body in consideration of the curing process of the resin was formulated. This simple formula expresses the storage elastic modulus and linear expansion coefficient of a resin that exhibits so-called viscoelastic behavior that changes significantly with time and temperature with a simple viscoelastic dynamics model, and is easily predicted from the basic formula of conventional strength of materials. It is an expression.

2) When the curing temperature obtained by using the photoelastic device was applied to the derived simple formula, it was found that the simple value was almost the same as the experimental value and the theoretical value. However, the content of the study here was verified for a two-layer laminate consisting of an epoxy resin and a printed circuit board, and the effects of the multilayer laminate and the heating and cooling rates have not been investigated. Therefore, the scope of application of this simplified formula will be an issue for the future, and further studies are expected.

15

Effect of Cooling Rate on Warpage Behavior of Laminates

15.1 Warp Deformation Experiment

15.1.1 Laminated Body Shape

Figure 15.1 shows a laminated body consisting of an epoxy resin and a printed circuit board that exhibit viscoelastic behavior as a simplified model of electronic components. Here, the epoxy resin was cured at room temperature at a ratio of 5:1 between the main agent and the curing agent and then heat-cured at 180°C for 30 minutes. The shape of the laminate was 150 mm in length and 5 mm in width, the thickness of the printed circuit board was constant at 1.0 mm, and the thicknesses of the epoxy resin were 2.0, 5.0, and 8.0 mm in three patterns.

15.1.2 Experimental Method

The laminate was heated and held from about 25°C to 180°C, which is a glass transition temperature of T_g or higher, where the thermal and mechanical properties of the epoxy resin and the printed circuit board change, and then cooled to about 25°C. Here, the cooling method is (i) When the laminate is quickly put into water from 180°C and rapidly cooled. (ii) When the laminate is cooled (air-cooled) by opening the door of the furnace.

There are two types. At this time, the temperature that changed from heating to cooling was measured by a thermocouple attached to the center of the epoxy resin strip, and the warp deformation behavior was sequentially photographed with a CCD camera, and the warp deformation amount was measured from this.

Polymeric Materials for Electronic Packaging, First Edition. Shozo Nakamura.

Published 2023 by John Wiley & Sons, Inc.

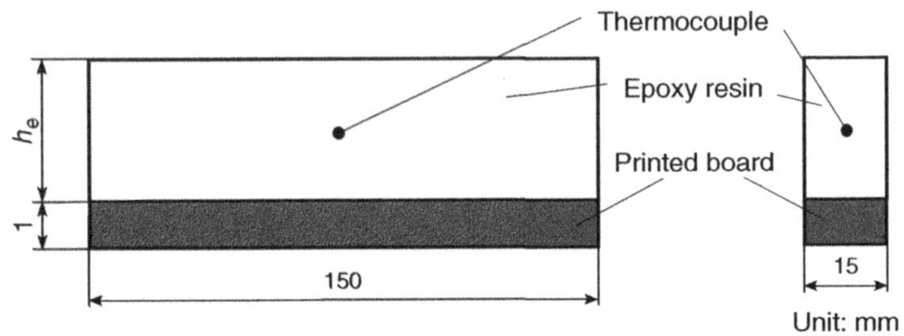

Figure 15.1 Shape and dimensions of the laminated structure.

15.2 VESAP Analysis

Table 15.1 shows the thermal and mechanical properties of the constituent materials of the laminate.

Figure 15.2 shows the linear expansion coefficients of the epoxy resin and the printed circuit board, respectively, and Fig. 15.3 shows the master curves of the relaxation elastic modulus of both materials.

The coefficient of linear expansion coefficient α of the epoxy resin and the printed circuit board was calculated based on the measured values of the temperature change of the thermal strain using the thermomechanical analyzer.

Table 15.1 Properties of the constituent materials.

Materials \ Properties		Thermal conductivity λ (W/m K)	Specific heat C (kJ/kg K)	Density ρ (kg/m^3)	Glass transition temperature T_g (°C)	Thermal expansion coefficient α (10^{-6}/K)		Modulus E (GPa)	
						$T \leqq T_g$	$T_g < T$	$T \leqq T_g$	$T_g < T$
Epoxy resin	Rapid	0.186	1.423	1220	100.6	45.5	154.7	2.09 ~ 0.03	
	Air				103.6	78.0	187.6	2.16 ~ 0.03	
	Heat				92.9	95.2	158.0	1.79 ~ 0.02	
Printed board		0.320	1.256	1810	119.2	14.5	12.2	9.73 ~ 1.80	

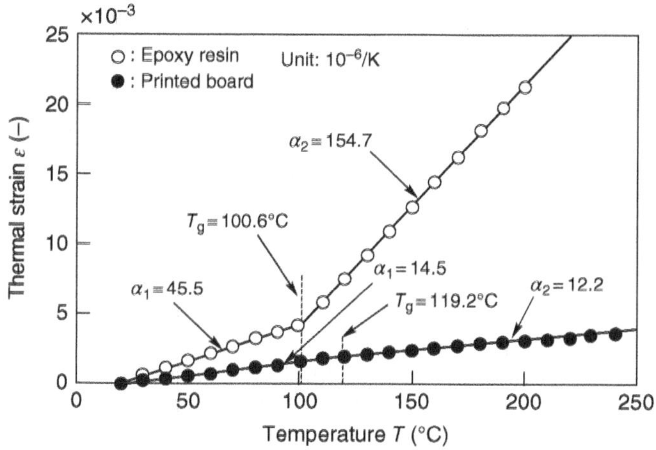

Figure 15.2 Thermal expansion coefficient for materials.

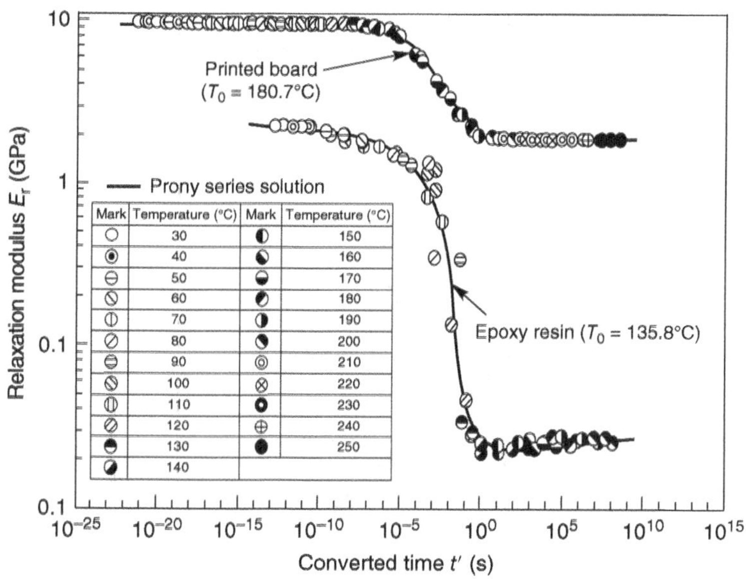

Figure 15.3 Master curve of relaxation modulus for epoxy resin and printed board.

On the other hand, the relaxation elastic modulus E_r was measured by changing the frequency to 1, 10, and 100 Hz at each desired temperature using the aforementioned dynamic viscous elastic modulus.

Next, these measured values are plotted on the converted logarithmic time axis under the reference temperature T_0.

As can be seen from Fig. 15.2, the T_g of the epoxy resin and the printed circuit board are 100.6°C and 119.2°C, respectively, and the coefficient of linear expansion differs between the low-temperature part (α_1) and the high-temperature part (α_2) with T_g as the boundary, which is peculiar to viscoelastic materials. Shows the nature. This epoxy resin showed values of $\alpha_1 = 45.5 \times 10^{-6}$/K and $\alpha_2 = 154.7 \times 10^{-6}$/K. The printed circuit board showed values of $\alpha_1 = 14.5 \times 10^{-6}$/K and $\alpha_2 = 15.2 \times 10^{-6}$/K.

On the other hand, as shown in Fig. 15.3, the relaxation elastic modulus E_r of the epoxy resin and the printed circuit board could be displayed with a single smooth master curve from an extremely short time to an extremely long time. This was approximated by the Prony series and used for VESAP analysis.

Figure 15.4 shows the time–temperature shift factor in creating the master curve in Fig. 15.3. As can be seen from Fig. 15.4, the epoxy resin and the printed circuit board could be approximated by two Arrhenius equations with different activation energies ΔH in the low-temperature part and the high-temperature part.

From this, it can be seen that the epoxy resin and the printed circuit board used here are materials to which the time–temperature conversion rule holds and the aforementioned linear viscoelasticity theory can be applied.

Various material property values being described earlier, VESAP analysis was performed on the warp deformation behavior of the laminate with different cooling conditions. Epoxy resin thickness was changed in three patterns as in the

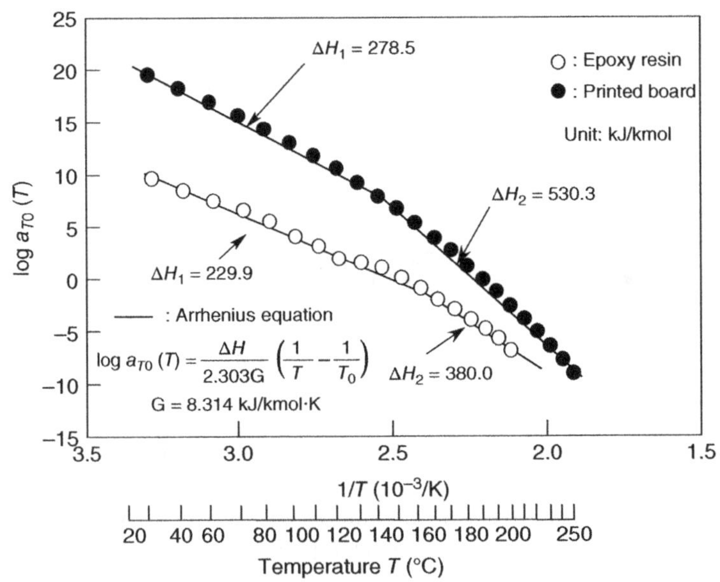

Figure 15.4 Time–temperature shift factor for epoxy resin and printed board.

experiment, and thermal analysis and stress analysis were kneaded. It was two types of temperature conditions were analyzed: a heating process from 25°C to 180°C, quenching from 180°C to 25°C, and air cooling.

15.3 Results and Considerations

15.3.1 Heat Transfer Coefficient

The heat transfer coefficient was determined by using a thermocouple attached to the center of the aforementioned epoxy resin strip in correspondence with the experimental value of the time change of temperature.

As an example, Fig. 15.5 shows the results of temporal changes in the temperature of quenching at an epoxy resin thickness of 5 mm.

Here, the plot is the experimental value, and the solid line is the result of thermal viscoelastic numerical analysis. When the heat transfer coefficient was calculated from Fig. 15.5, the value was 267 W/(m^2 K). The heat transfer coefficient under other conditions is also obtained by the same method, and these values are summarized in Table 15.2.

15.3.2 Warp Deformation Behavior

Figure 15.6a–c shows the warp deformation behavior when a series of heat loads from heating to cooling are applied to the laminate for various epoxy resin thicknesses.

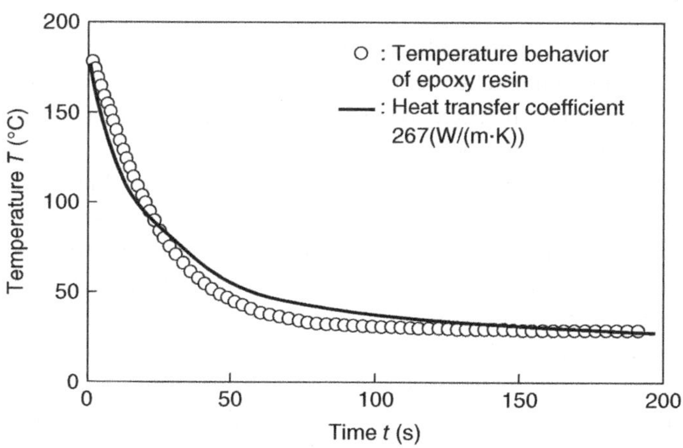

Figure 15.5 Temperature changes and heat transfer coefficient.

Table 15.2 Heat transfer coefficient.

Unit: (W/m^2 K)

Heat load condition \ Thickness of epoxy resin (mm)	2	5	8
Rapid	291	267	233
Air	10	10	9
Heat	15	15	14

That is, the experimental and theoretical values are compared and shown for a series of heat loads in the heating process from room temperature of about 25°C to 180°C and the subsequent cooling process from 180°C to 25°C (air cooling and quenching).

Here, the plots in Fig. 15.6 show the experimental values of air cooling and quenching. The solid line is the theoretical value for air cooling, the broken line is the theoretical value for quenching, and the alternate long and short dash line is the theoretical value for heating.

The warp deformation behavior during the heating process is shown on the left side of Fig. 15.6a–c. As for the warp deformation behavior of the laminate, the epoxy resin side begins to warp in a convex state with heating regardless of the epoxy resin thickness, and the warp deformation amount tends to increase as the temperature rises.

The amount of warpage deformation changes its gradient in the temperature range of 80°C to 140°C, and then reaches its maximum value at 180°C. The elastic modulus and linear expansion of the epoxy resin and the printed circuit board are the causes of such warp deformation behavior.

In the case of the laminate having an epoxy resin thickness of 2 mm in Fig. 15.6a, as the temperature rises, the epoxy resin side begins to warp in a convex state, and the amount of warping also increases. However, in the temperature range of 80°C to 140°C, the amount of warpage becomes moderate or even tends to decrease.

I presume that this phenomenon is caused by the balance between the expansion force on the epoxy resin side and the force on the printed circuit board side. In other words, in the range of 90°C to 140°C, the elastic modulus of the epoxy resin decreases significantly (2.09 → 0.03 GPa), which corresponds to the glass transition temperature of the epoxy resin. So, the thermal expansion force of the epoxy resin decreases significantly.

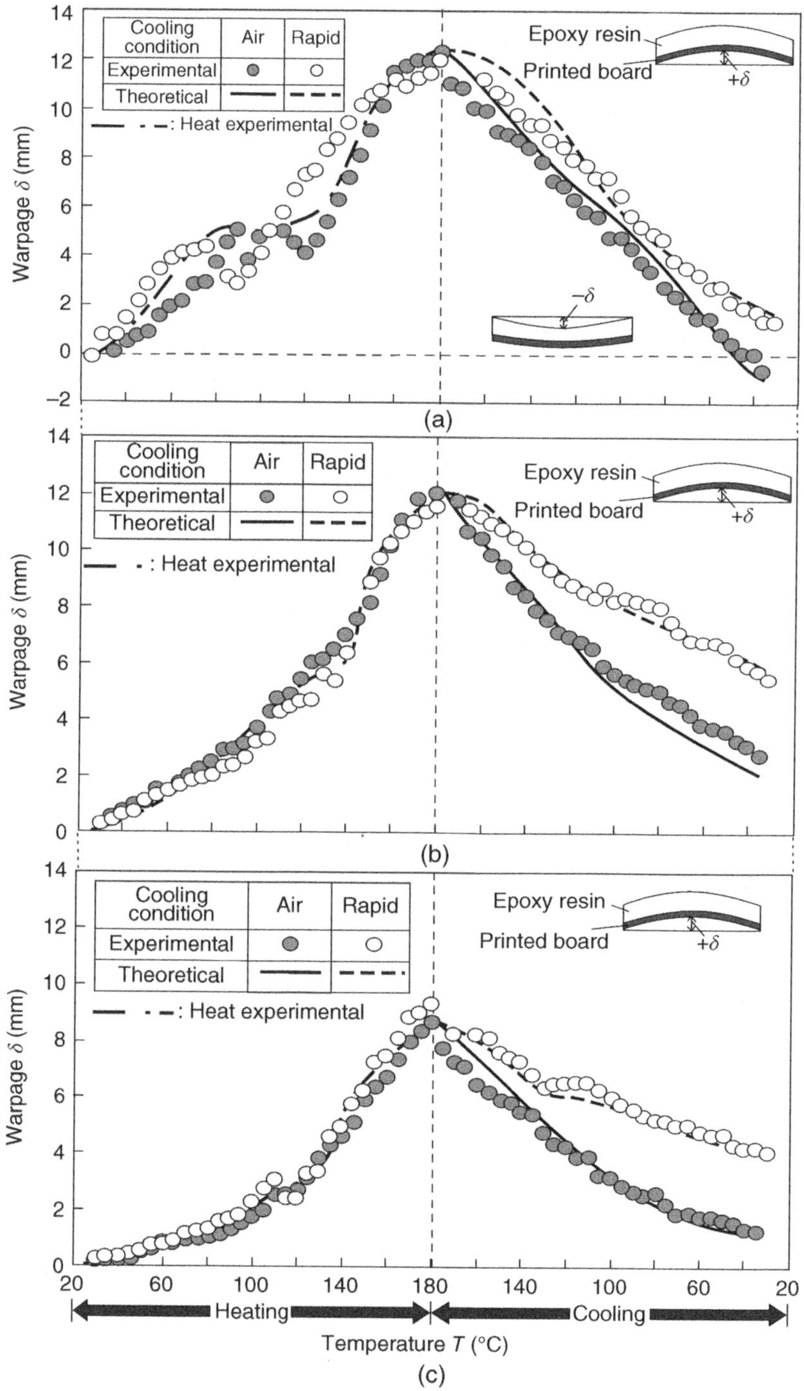

Figure 15.6 Warpage behavior of laminated structure by thermal load. (a) $h_e = 2$ mm, (b) $h_e = 5$ mm, and (c) $h_e = 8$ mm.

On the other hand, since the printed circuit board has a large elastic modulus, a force acts to restore the warp deformation. As a result, the warp deformation amount of the laminate becomes moderate or tends to decrease.

When the elastic modulus of the printed circuit board decreases (9.73 GPa → 1.80 GPa), it is considered that it increases again due to the thermal expansion force of the epoxy resin.

On the other hand, in the case of Fig. 15.6b,c, the gradient of the amount of warp deformation in the temperature range between 80°C and 140°C does not become gentle. It is considered that this is because the thickness of the epoxy resin is so thick that the force to return the printed circuit board to the original state is less affected.

The warp deformation behavior during the cooling process is shown on the right side of Fig. 15.6a–c. In either case, the amount of warpage deformation is reduced because the epoxy resin shrinks as a result of cooling. When comparing air cooling and quenching here, the amount of warpage deformation is smaller in air cooling than in quenching. In addition, only in the case of rapid cooling with an epoxy resin thickness of 8 mm, a change in the gradient of the amount of warpage deformation was observed near 140°C.

Furthermore, as can be seen from these series of figures, the results of the thermal viscoelastic numerical analysis are in good agreement with the experimental values. Therefore, the warp deformation behavior can be predicted by thermoviscoelastic number analysis under both air-cooling and quenching conditions.

15.4 Final Warp Deformation and Residual Warp Deformation

Figure 15.7 shows the relationship between the epoxy resin thickness and the final warp deformation amount δ_f at the final stage (at 25°C) when a series of heat loads from heating to cooling are applied to the laminate. As can be seen from Fig. 15.7, the tendencies of the final warp deformation amount of quenching and air cooling are the same.

In addition, when the epoxy resin is air-cooled at a thickness of 2 mm, the printed circuit board side becomes convex and a reverse warp phenomenon is observed.

On the other hand, Fig. 15.8 shows the residual warp deformation amount and the epoxy resin thickness when the warp deformation amount at 180°C is set to 0.

As can be seen from Fig. 15.8, the tendencies of the residual warp deformation amount of quenching and air cooling are also the same.

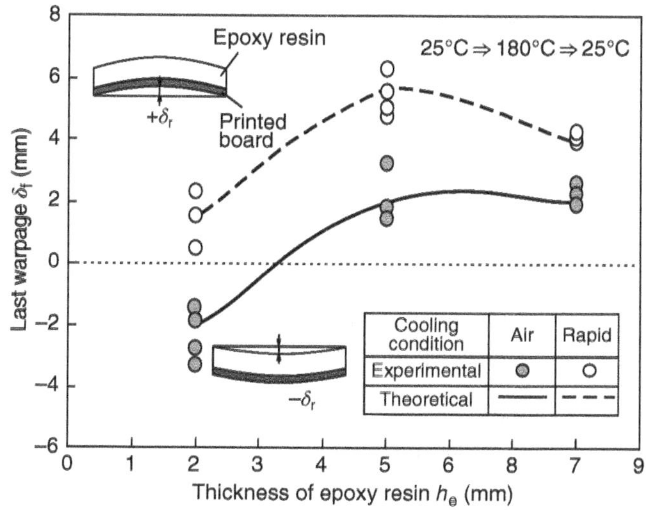

Figure 15.7 Relationship between epoxy resin thickness and final warp deformation amount.

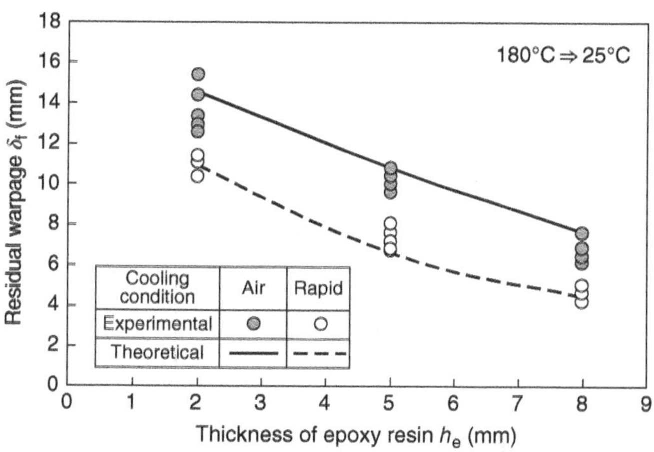

Figure 15.8 Residual warp deformation and epoxy resin thickness.

Furthermore, the amount of residual warpage deformation in any epoxy resin thickness is larger in air cooling, and the difference in the amount of residual warp deformation between air cooling and quenching is almost constant.

However, the cause and mechanism of a large amount of residual warpage deformation in the case of air cooling have not been clarified. At present, we suspect that this is due to the flexural rigidity of the laminate.

15.5 Estimation Mechanism Between Cooling Rate and Deformation of Laminate

In Sections 15.3 and 15.4, the warpage deformation behavior of the two-layer laminate was investigated for rapid cooling and slow cooling.

In this section, I discuss how the cooling rate, such as quenching and gradual cooling, affects the warp deformation of laminates. speculatively inferred the relationship between laminate temperature changes and the viscoelastic mechanical model of springs, dashpots, and thermal expansion.

Figure 15.9a shows a model of a two-layer laminate model, and (b) shows a viscoelastic mechanical model consisting of springs, dashpots, and thermal expansion.

In Fig. 15.9, the viscoelastic model is representatively shown for the central part (X, Y parts) of each material (Figs. 15.10 and 15.11).

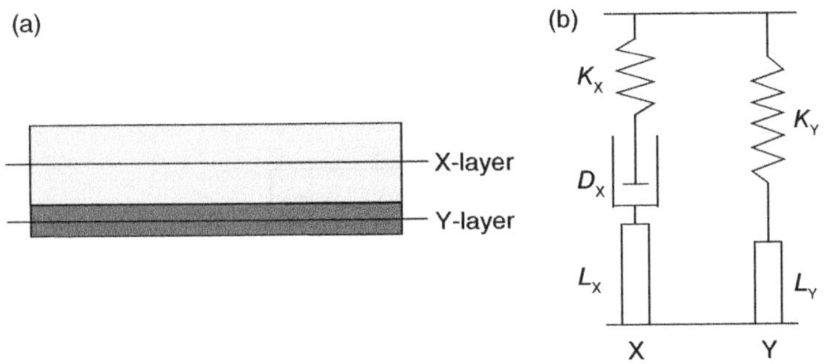

Figure 15.9 Laminated model and viscoelastic model. (a) Two-layer model and (b) mechanical model.

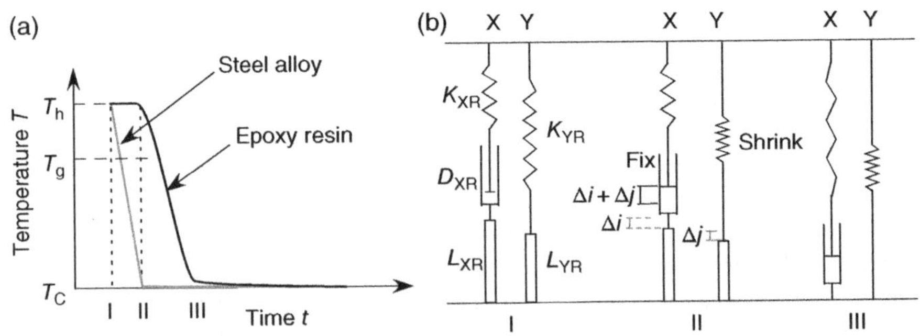

Figure 15.10 Changes of temperature and viscoelastic model in case of rapid cooling. (a) Temperature changes of the two-layer model. (b) Residual warpage mechanism by mechanical model.

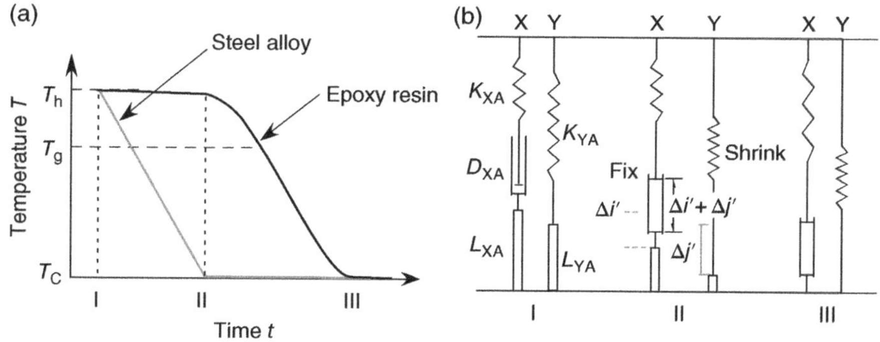

Figure 15.11 Changes of temperature and viscoelastic model in case of slow cooling. (a) Temperature changes of the two-layer model. (b) Residual warpage mechanism by viscoelastic model.

15.6 Conclusion

Using a laminate consisting of an epoxy resin and a printed substrate, a series of heat loads from heating to cooling are applied to it, and the warp deformation behavior when the cooling rate is variously changed is obtained by experiments, and thermal viscoelastic numerical analysis is performed.

The following conclusions were obtained by comparing them with the results.

1) By performing thermal viscoelastic numerical analysis, it is possible to predict the warp deformation behavior regardless of the cooling conditions.
2) The amount of residual warpage deformation during cooling is larger in air cooling than in quenching at any epoxy resin thickness, and the difference is almost constant.

Appendix A

Development of Viscoelastic Analysis Software (VESAP)

A.1 Development Needs and Concepts

The motives and backgrounds that the author strongly felt for the need for VESAP (Viscoelastic Stress Analysis Program) development, as well as the aims of the development, are summarized as follows.

1) There are many analysis software related to strength and deformation in the world, but most of them are based on elastic theory. These analysis software are not suitable for viscoelastic materials whose physical properties are greatly affected by temperature and time, such as resins.
2) Preparation for VESAP analysis should be as simple and easy as possible, and at the same time, it should be easy for engineers and researchers of companies to operate even if they are not experts in viscoelasticity.
3) VESAP software can be applied not only to LSI packages composed of a wide variety of materials including viscoelastic materials such as resin but also to structures such as general laminated substrates.
4) The analysis results by VESAP are reliable competitions that are in good agreement with the experimental values. In other words, the absolute values of the analysis results such as stress, strain, and warpage deformation amount generated by the applied thermal load are in good agreement with the experimental values.

From the aforementioned background, the viscoelastic analysis software (VESAP) developed by the author based on the issues described so far will be explained as shown in Fig. A.2.

One of the features as mentioned earlier is that the preparation for viscoelasticity analysis has been simplified. The focus is on the ability of engineers to

Polymeric Materials for Electronic Packaging, First Edition. Shozo Nakamura.

Published 2023 by John Wiley & Sons, Inc.

avoid complicated analysis preparation stages as much as possible and realize practical design.

Currently, many analysis software such as AVAQUS, ANSYS, and MARC are on the market. When using these software, it is often difficult to select the elements required by the finite element method (FEM), divide the mesh, and input the viscoelastic properties of the material. The preparatory steps such as plane strain/ plane stress, three-dimensional analysis, and coupled analysis of heat and stress are complicated and tend to take much time to master.

In addition, there is a problem in terms of accuracy, such as a large discrepancy between the actual product and the analysis result.

The VESAP developed by the author based on these experiences is simple and can be used relatively easily. The analysis time of VESAP is also related to the object to be analyzed, but it is extremely short, generally from several seconds to several minutes.

① Temporal change in temperature
② Temporal change in curvature
③ Temporal change in the amount of warpage deformation
④ Thermal stress
⑤ Time change of strain can be simulated quickly and at the same time

From here, we will derive the basic formula of VESAP analysis and verify the validity of the analysis result by giving various cases such as multi-layered laminated structures.

The results of both VESAP analysis and experiments are mainly focused on the warp deformation behavior corresponding to a series of heat loads from heating to cooling, which is a problem in resin adhesion to large area silicon wafers such as WLP, which has been attracting attention these days.

I will also introduce examples that can be applied to various LSI package fields such as CSP-μBGA. In particular, CSP-BGA is widely used as an LSI package in recent electronic equipment, and it has a particularly thin structure and various resins are used, and it is important to ensure its thermal deformation and strength reliability.

A.2 Derivation of Basic Formula for Analysis

This section explains the calculation used for viscoelasticity analysis. Consider a laminate having a rectangular cross section consisting of n layers of materials 1, 2, 3, ..., n as shown in Fig. A.1.

Let l be the length of the laminate, d be the depth, and h be the thickness, and it is assumed that there is a relationship of $l \gg d \gg h$ between them. Cartesian coordinates with x in the depth direction, y in the thickness direction, and z in the

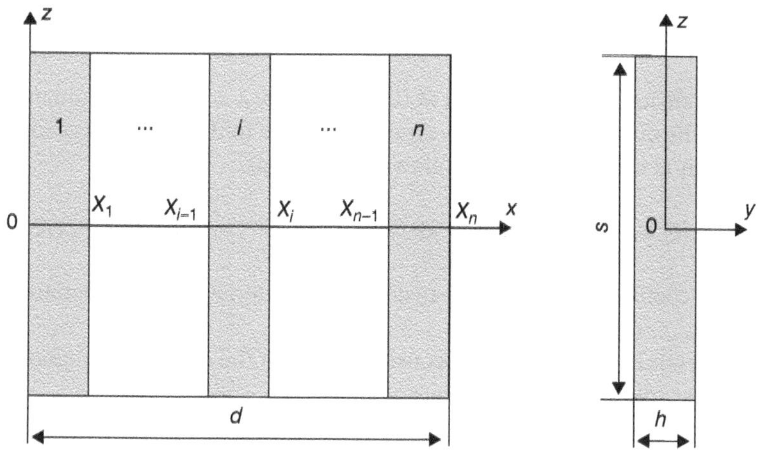

Figure A.1 Laminated model consisting of *n* layers.

longitudinal direction are provided; the surface is $x = 0$, the interface between material 1 and material 2 is $x = x_1$, and so on. Let $x = x_3, \ldots, x = x_n$.

Consider the case where the entire laminate is slowly cooled from the holding temperature T_h or both surfaces ($x = 0$, $x = d$) are rapidly cooled to the cooling temperature T_c. In the former case of slow cooling, the temperature inside the laminate is uniform, and the function $T(t)$ is only for the time t.

As for the temperature distribution in quenching, if both sides ($y = \pm h/2$) of the beam are insulated and the temperature conductivity of the constituent material is constant without changing with temperature, it is unsteady generally used in thermodynamics. It can be obtained from the equation of heat conduction, and the temperature distribution is a function $T(x, t)$ of the coordinates x and the time t.

The thermal stress is generated inside the laminate only in the z-axis direction due to the temperature distribution $T(x, t)$ that changes over time. This thermal stress in the z-axis direction is a function of coordinates x and time t, and this is $\sigma(x, t)$. Assuming that the strain due to this thermal stress is $\varepsilon(x, t)$, the constituent material is a linear thermoviscoelastic body between the thermal stress $\sigma(x, t)$ and the strain $\varepsilon(x, t)$, and if time–temperature conversion rule holds, then the relationship of the following equation holds.

$$\sigma(x,t) = \int_0^t E_{ri}(t' - \tau', T_0) \frac{d}{d\tau'} \varepsilon(x, \tau') d\tau' \tag{A.1}$$

where E_{ri} (t', T_0) is the relaxed elastic modulus of the i (1, 2, ..., n) layer at the reference temperature T_0, and t', τ' is the conversion time, which is the time of the i layer. If the temperature shift factor is $a_{i_{T_0}}(T)$, it can be obtained from the following equation.

$$t' = \int_0^t \frac{du}{a_{i_{T_0}}\{T(x,u)\}}, \quad i = 1 \sim n \tag{A.2}$$

Next, if the expansion and contraction at the coordinate $x = 0$ of the laminated body is $\varepsilon_n(t)$ and the curvature is $\kappa(t)$, the strain $\varepsilon(x, t)$ due to the stress in the longitudinal direction is expressed by the following equation.

$$\varepsilon(x,t) = \varepsilon_n(t) + \kappa(t)x - \int_{T_h}^{T(x,t)} \alpha_i(T)dt \quad i = 1 \sim n \tag{A.3}$$

where $\alpha_i(T)$ is the coefficient of linear expansion of the material at the temperature T of the i ($i = 1,2, \ldots, n$) layer.

Now, since this laminated body is not constrained from the outside throughout the cooling process, the following force

$$\int_0^d \sigma(x,t)dx = 0 \tag{A.4}$$

balance equation and moment balance equation are established.

$$\int_0^d \sigma(x,t)x\,dx = 0 \tag{A.5}$$

Therefore, by substituting Eqs. (A.1)–(A.3) into Eqs. (A.4) and (A.5), the following differential integral equation can be obtained.

$$\sum_{i=1}^{n} \int_{x_{i-1}}^{x_i} \int_0^t E_{ri}(t' - \tau', T_0) \frac{d}{d\tau'} \left\{ \varepsilon_n(\tau') + \kappa(\tau')x - \int_{T_h}^{T(x,\tau')} \alpha_i(T)dT \right\} d\tau'$$
$$dx = 0 \quad i = 1 \sim n \tag{A.6}$$

$$\sum_{i=1}^{n} \int_{x_{i-1}}^{x_i} \int_0^t \left[E_{ri}(t' - \tau', T_0) \frac{d}{d\tau'} \left\{ \varepsilon_n(\tau') + \kappa(\tau')x - \int_{T_h}^{T(x,\tau')} \alpha_i(T)dT \right\} \right] x\,d\tau'$$
$$dx = 0, i = 1 \sim n \tag{A.7}$$

Equations (A.6) and (A.7) are basic equations for obtaining the thermal stress and the amount of warpage deformation that occur when the laminate is cooled regardless of whether it is slowly cooled or rapidly cooled. That is, the expansion and contraction $\varepsilon_n(t)$ and the curvature $\kappa(t)$ can be obtained by solving Eqs. (A.6) and (A.7) as simultaneous equations.

In addition, the thermal stress distribution $\sigma(x, t)$ can be obtained from Eqs. (A.1) and (A.2). Then, $\sigma(x, t)$, $\varepsilon_n(t)$, and $\kappa(t)$ remain after sufficient time has passed after quenching, and the entire laminate remains uniform at the cooling temperature T_c is the residual thermal stress $\sigma_r(x)$, the residual expansion and contraction ε_{nr}, and the residual curvature κ_r.

The thermal stress and curvature induced in this way can be quantitatively calculated by substituting the physical properties of the resin materials that make up various semiconductor packages and electronic components used in electronic devices, which will be described later.

These numerical values contribute to the optimization of material design, structural design and process design of various electronics equipment with highly reliability.

A.3 Contents of the Developed VESAP Software

As described earlier, the calculation formula that is the basis of VESAP analysis was derived. Then, these basic formulas were programmed. And, these basic equations have universality that can be applied to any laminated body composed of multiple layers.

For example, the elastic modulus changes greatly with temperature, and this is approximated by the Prony series under the reference temperature using the time–temperature conversion rule.

In addition, the time–temperature shift factor at that time is approximated by the Arrhenius equation and substituted into the calculation equation.

Here, a series of these contents can be easily operated on a PC. The outline is schematically shown in Fig. A.2.

Also, as an example, a VESAP analysis results of the temperature distribution inside the laminate, which changes over time, in the case of air cooling, rapid cooling, and a combination of cooling from both surfaces for a laminate of Al and epoxy resin shown in Fig. A.3.

From Fig. A.3, it can be seen at a glance that the temperature change and the temperature distribution inside the laminate change depending on the cooling conditions.

For reference, Fig. A.4 shows an overview of the flow of VESAP analysis.

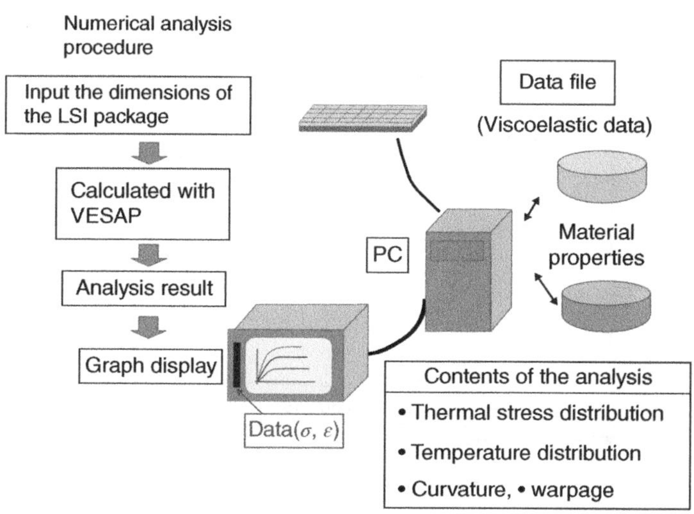

Figure A.2 VESAP analysis method and conceptual diagram of result display.

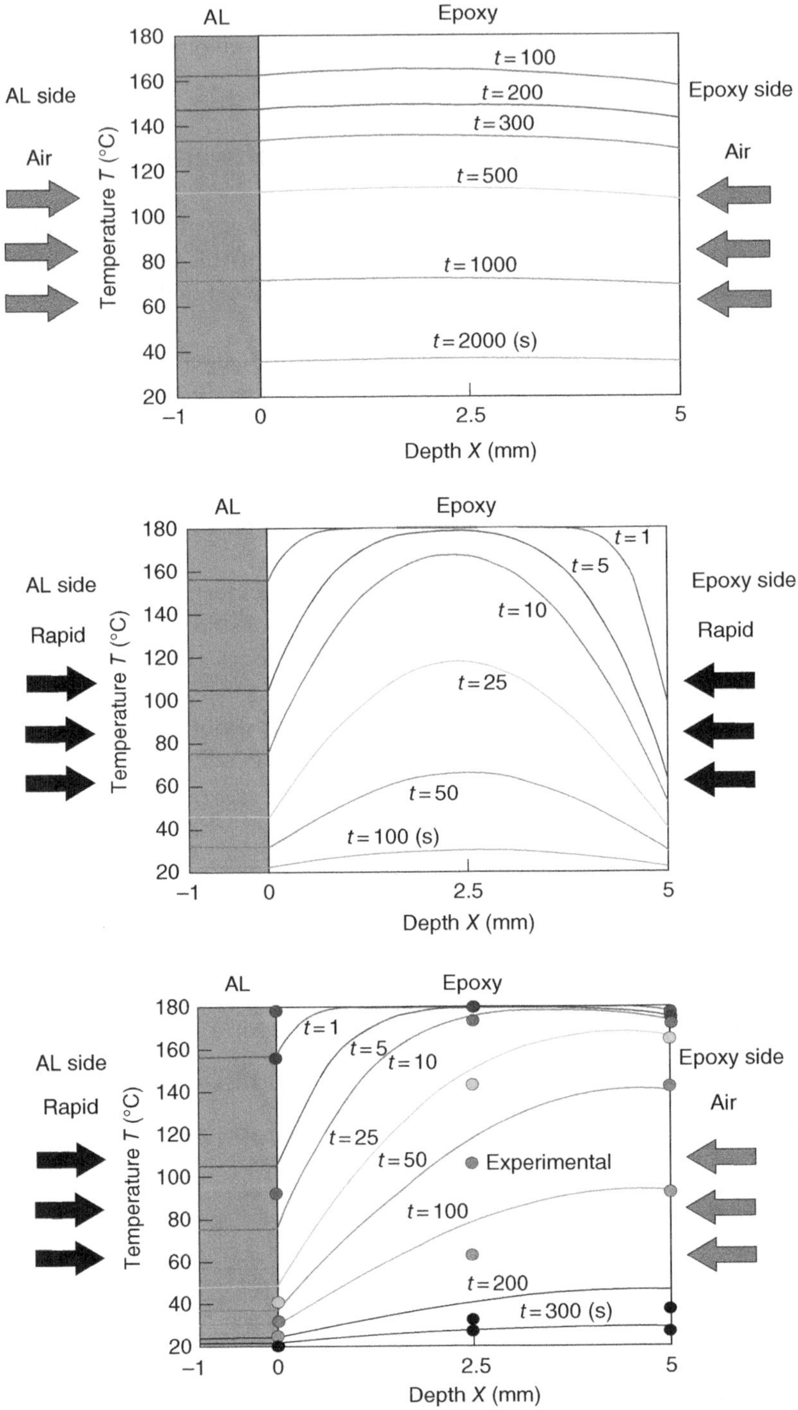

Figure A.3 Time-varying temperature distribution inside the laminate.

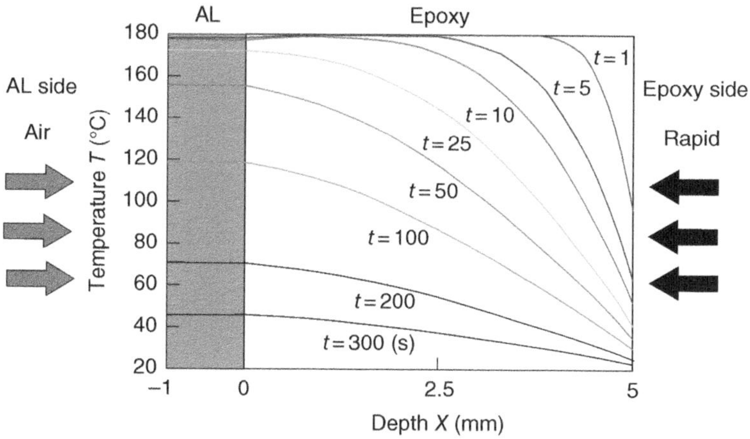

Figure A.3 (Continued)

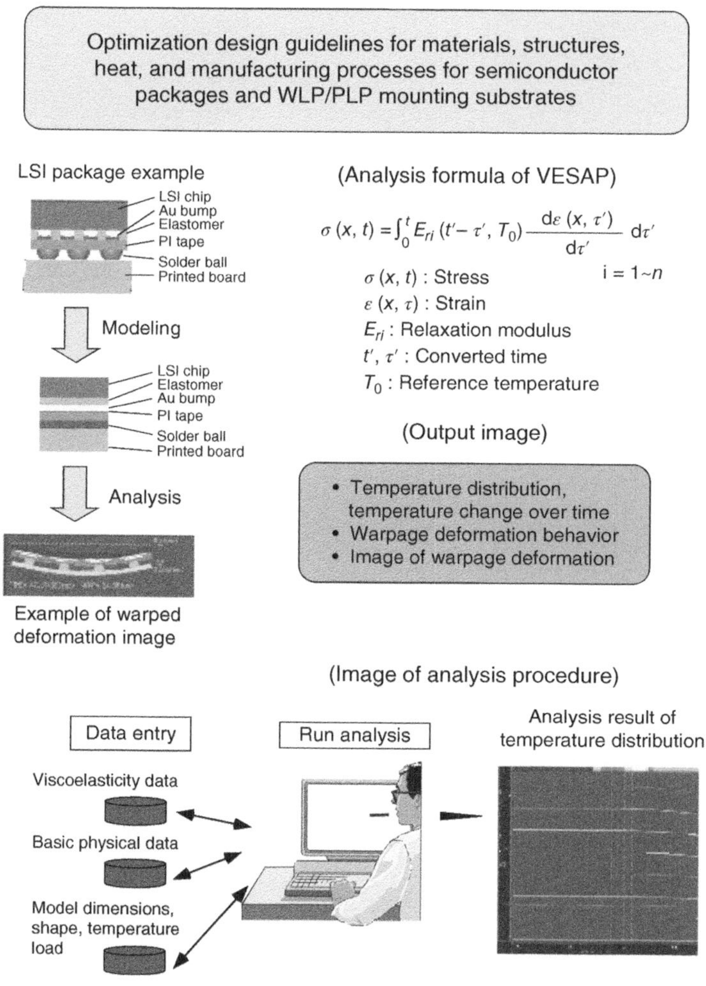

Figure A.4 VESAP analysis overview.

Bibliography

Theory of Elasticity and Viscoelasticity

Murakami, G. (1998). Special issue on semiconductor package mounting technology and future trends in mounting. *Journal of the Institute of Electronics Packaging*, 4(1): 2–4.

Murakami, G. (2007). *Illustrated All About Cutting-Edge Semiconductor Packaging Technology* (ed. Semiconductor New Technology Study Group). Industrial Research Institute.

Murakami, G. (1988). *Lead Frame of LOC, COL Structure for Inserting 100 mm^2 Chip into 300 mil DIP. Nikkei Microdevices.*

Kasuga, H. (1998). Surface mount pocket book Ultra small package CSP/BGA technology, Edited and Written Nikkan Kogyo Shimbun.

Timoshenko, S. and Goodier, N. (1973). *Theory of Elasticity*, 3e. McGraw Hill.

Hideyoshi and Udoguchi (1957). *Elasticity.* Kyoritsu Shuppan.

Maezawa, S. (1970). *Elasticity.* Morikita Publishing.

Sakata, M. (1970). *Solid Mechanics.* Asakura Shoten.

Tamateto (1971). *Deformation of Elastic Bodies.* Corona Publishing.

Love, A.E.H. (1927). *A Treatise on the Mathematical Theory of Elasticity*, 4e. Cambridge University Press. Reprinted by Dover Publications, 1963.

Green, A.E. and Zerna, W. (1968). *Theoretical Elasticity*, 2e. Oxford University Press.

Mogami, T. (1942). *Two-Dimensional Elasticity.* Corona Publishing.

England, A.H. (1971). *Complex Variable Methods in Elasticity.* Wiley.

Miyamoto, H. (1967). *Three-Dimensional Elasticity.* Shokabo.

Ishida, M. (1976). *Elastic Analysis of Cracks and Stress Intensity Factors.* Baifukan.

Okamura, H. (1976). *Introduction to Linear Fracture Mechanics.* Baifukan.

Williams, J.G. (1973). *Stress Analysis of Polymer.* Longman.

Kunio, T. and Miyano, Y. (1970). Optical viscoelasticity experimental method. *Non-Destructive Inspection* 19.

Nakagawa, T. (1960). *Rheology.* Iwanami Shoten.

Polymeric Materials for Electronic Packaging, First Edition. Shozo Nakamura.

Published 2023 by John Wiley & Sons, Inc.

Goto, R., Hirai, N., and Hanai, T. (1962). *Rheology and Its Applications*. Kyoritsu Shuppan.
Oka, K. (1955). *Viscoelasticity of Polymers*. Iwanami Koza.
Ferry, J.D. (1960). *Viscoelastic Properties of Polymer*. Wiley.

Things Related to Semiconductor Packages and Electronic Devices

Murakami, G., Mita, M., and Nakamura, S. (1998). Achieving an optimum structural package design with thermo-viscoelastic analysis. *Chip Scale Review* 2 (5).
Nakamura, S., Murakami, G., and Isaka, K. (1999). Prediction of warpage deformation behavior by thermal viscoelastic analysis of semiconductor device by FCA method. *Journal of Electronics Packaging Society* 2 (4).
Murakami, G., Nakamura, S., and Mita, M. (2000). Optimization design of elastomer structure of CSP-μBGA by thermoviscoelastic analysis. *Journal of Electronics Packaging Society* 2 (4).
Kawanobe, T. and Murakami, H. (1985). "IC" Special Issue on Child Parts, Electronic Materials.
Naito, T., Murakami, G., and Taketani, N. (1988). *1 mm-Thick LSI Package Breaking the Image of "Thin and Weak"*. Nikkei Microdevices.
Murakami, G. (1988). *Lead Frame of LOC, COL Structure for Inserting 100 mm^2 Chip into 300 mil DIP. Nikkei Microdevices.*
Murakami, G. (1991). *Ideas that Change the Times – QFP*. Nikkei Microdevices.
Murakami, G. (1998). Special issue on semiconductor package mounting technology and future trends in mounting. *Journal of the Institute of Electronics Packaging*, First Issue 1 (1).

General Viscoelastic Relations

Yamamoto, K. (1966). Internal in metal-epoxy resin composites. *Analysis of Partial Stress, Nitto Giho* 41 (11).
Hijikata, F. (1973). Internal stress of epoxy resin cast products. *Mitsubishi Electric Technical Review* 47 (10).
Yamaoka, U. (1980). Curing shrinkage and internal stress of epoxy resin. *Materials* 29 (323).
Shimbo, Y. (1983). Internal stress of filled epoxy resin. *Kobunshi Ronbunshu* 40 (1).
Oga, T. (1983). Part stress. *Material Testing Techniques* 28 (1).
So, P. and Broutmant, L.J. (1976). Residual stresses in polymers and their effect on mechanical behavior. *Polymer Engineering and Science* 16 (12).
Miyano, Y. and Tamura, T.H. (1968). Determination of photoviscoelastic characteristic coefficient of polyurethane resin and its calculation method. *Transactions of the Japan Society of Mechanical Engineers* 34 (262).

Miyano, Y., Tamura, T., and Kunio, T. (1969). The mechanical and optical characterization of polyurethane with application to photoviscoelastic analysis. *Bulletin of JSME* 12 (49).

Miyano, Y., Shimbo, M., and Kunio, T. (1980). Elucidation of residual stress generation mechanism by quenching of epoxy resin beams. *Transactions of the Japan Society of Mechanical Engineers* 46 (407).

Miyano, Y., Shimbo, M., and Kunio, T. (1981). Elucidation of residual stress generation mechanism by quenching of epoxy resin beams. *Transactions of the Japan Society of Mechanical Engineers* 47 (419).

Sugimori, S., Miyano, Y., and Kunio, T. (1983). Double-sided quenching of epoxy resin perforated strip photo-viscoelastic analysis of thermal stress by using. *Transactions of the Japan Society of Mechanical Engineers* 50 (458).

Miyano, Y., Nakamura, S., and Kunio, T. (1985). Proposal and theoretical study of simple optical measurement method by quenching thermosetting resin strip. *Transactions of the Japan Society of Mechanical Engineers* 51 (463).

Morland, L.W. and Lee, E.H. (1960). Stress analysis for linear viscoelastic materials with temperature variation. *Transactions of the Society of Rheology* 4.

Williams, M.L., Landel, R.F., and Ferry, J.D. (1955). The temperature dependence of relaxation mechanisms in amorphous polymers and other glass-forming liquids. *Journal of the American Chemical Society* 77.

Muki, R. and Sternberg, E. (1961). On transient thermal stress in viscoelastic materials with temperature dependent properties. *Journal of Applied Mechanics* 28.

Mugi (1961). Recent progress in linear viscoelasticity theory. *Transactions of the Japan Society of Mechanical Engineers* 64 (513).

Lee, E.H. and Rogers, T.G. (1963). Solution of viscoelastic stress analysis problems using measured creep or relaxation function. *Journal of Applied Mechanics* 30.

Lee, E.H., Rogers, T.G., and Woo, T.C. (1965). Residual stress in a glass plate cooled symmetrically from both surfaces. *Journal of the American Ceramic Society* 48 (9).

Woo, T.C., Jones, J.W., and Ting, T.C.T. (1971). Transient and residual thermal stress in a viscoelastic cylinder. *Developments in Mechanics: Vol. 6, Proceedings of the 12th Midwestern Mechanics Conference Held at the University of Notre Dame* (16–18 August 1971). Notre Dame: University of Notre Dame Press.

Frutiger, R.L. and Woo, T.C. (1971). A thermoviscoelastic analysis for circular plates of thermorheologically simple material. *Journal of Thermal Stresses* 12 (1).

Crochet, M.J. and Denayer, A. (1980). Transient and residual thermoviscoelastic stress in glass. *Journal of Applied Mechanics* 47.

Nakamura, S., Miyano, Y., and Kunio, T. (1982). Photoviscoelastic consideration of residual stress caused by rapid cooling of thermosetting resin. *Proceedings of the Japan Society of Mechanical Engineers*, No. 824-2.

Miyano, Y., Nakamura, S., and Sugimori, S. (1982). Experimental study of simple optical measurement method for residual stress generated in thermosetting resin beams hardened on both sides. *Materials Systems* 1.

Shibuya, Y. (1986). *Photoelastic Method as an Effective Means for Measuring Stress in Molds.* Nikkei Microdevices.

Gallagher, R.H. (1977). *Fundamentals of the Finite Element Method.* Maruzen.

Miyano, Y., Nakamura, S., and Sugimori, S. (1986). A simplified optical method for measuring residual stress by rapid cooling in thermosetting resin strip. *Experimental Mechanics* 26 (2).

Nakamura, S., Miyano, Y., and Kaneda, A. (1988). Thermoviscoelastic analysis of residual stresses in a thermosetting resin/metal laminated beam caused by cooling. *JSME International Journal, Series 1* 31 (1).

Nakamura, S. (1988). Research on theoretical and experimental analysis of residual stress in viscoelastic laminates. Doctoral Dissertation.

Nakamura, S., Miyano, Y., and Goto, M. (1990). Prediction of residual stress in LSI plastic package by thermal viscoelastic analysis of three-layer laminated beam. *Transactions of the Japan Society of Mechanical Engineers* 56 (525, A).

Nakamura, S., Tsunoda, S., and Kaneda, A. (1991). Low stress process for LSI packages. *Plasticity and Processing* 32 (369).

Nakamura, S. (1995). Warp deformation behavior of plastic composite substrates by thermal viscoelasticity analysis. *Electronics Mounting Technology Association, 6th MES Lectures.*

Nakamura, S. and Serizawa, K. (1997). Thermal viscoelastic analysis of warpage deformation behavior and residual stress of FCA mounting structure. *Journal of Circuit Packaging Society* 12 (3).

Nakamura, S., Kono, T., and Saeki, J. (1998). Prediction of warpage deformation behavior by viscoelastic analysis of laminate composed of resin and metal. *Molding Process* 10 (11).

Nakamura, S., Kushizaki, Y., and Goto, M. (2002). Thermoviscoelastic numerical analysis of influencing factors of material properties on residual stress in semiconductor devices. *Journal of the Institute of Electronics Packaging* 5 (4).

Nakamura, S., Goto, M., and Nakajima, Y. (2002). Thermoviscoelastic analysis of thermal load conditions on thermal residual stress and warpage deformation behavior of components of electronic components. *Journal of the Institute of Electronics Packaging* 5 (7).

Nakamura, S., Kushizaki, Y., and Goto, M. (2003). Optimization of layer configuration on thermal residual stress and warp deformation behavior of electronic components using thermoviscoelastic analysis. *Journal of Electronics Packaging Society* 6 (1).

Nakamura, S., Goto, M., and Kushizaki, Y. (2003). Thermal viscoelastic analysis of physical property influencing factors on thermal residual stress and warpage deformation behavior of plastic laminates. *Molding* 15 (4).

Nakamura, S., Goto, M., and Kushizaki, Y. (2004). Optimization of layer structure and physical properties of new BOC package by viscoelastic analysis. *Journal of Electronics Packaging Society* 7 (3).

Nakamura, S., Sakata, R., and Kushizaki, Y. (2004). Thermal viscoelastic properties in the curing process of adhesives for electronic parts. *Molding* 16 (6).

Nakamura, S. and Goto, M. (2005). Simple evaluation method for determining residual stress in viscoelastic laminates. *Transactions of the Japan Society of Mechanical Engineers, A* 71 (703).

Nakamura, S. and Senbo, Y. (2005). Thermal degradation behavior of polymer materials for ultra-thin semiconductor devices. *Molding* 17 (4).

Kushizaki, Y., Nakamura, S., and Kawabata, T. (2006). Prediction of thermal residual stress generated between wirings during heat cycle reliability evaluation using finite element analysis. *Molding Process* 18 (2).

Nakamura, S., Kawabata, T., and Fukui, Y. (2008). Simple evaluation method for determining warpage deformation behavior of viscoelastic laminated beams composed of epoxy resin and steel. *Materials* 57 (5).

Nakamura, S. and Kawabata, T. (2008). Simple evaluation method for determining warp deformation behavior of laminated beam composed of combination of viscoelastic materials. *Materials* 57 (9).

Nakamura, S., Tanaka, T., and Kawabata, T. (2008). Theoretical and experimental investigation of warpage deformation prediction of viscoelastic laminate accompanied by curing shrinkage and thermal shrinkage. *Materials* 57 (11).

Nakamura, S., Fukui, Y., and Yoshimura, K. (2008). Theoretical and experimental investigation of warpage deformation behavior of viscoelastic two-layer laminates under thermal load. *Journal of the Institute of Electronics Packaging* 11 (7).

Nakamura, S., Tanaka, T., and Katsuyama, Y. (2009). Thermoviscoelastic warp deformation analysis due to thermal load of laminate composed of thermally degraded epoxy resin and steel material. *Materials* 58 (9).

Nakamura, S., Tanaka, T., and Shinohara, T. (2010). Analysis of hardening shrinkage strain in the warp deformation process of viscoelastic laminated beams composed of thermosetting resin and steel. *Transactions of the Japan Society of Mechanical Engineers, A* 76 (768).

Nakamura, S., Suehiro, K., and Mishima, R. (2013). Experimental and theoretical investigation of warp deformation behavior of viscoelastic three-layer laminate under thermal load. *Journal of Electronics Packaging Society* 16 (1).

Nakamura, S., Suehiro, H., and Nakagawa, S. (2013). Simplified evaluation method of residual thermal warpage for viscoelastic two-layer-laminated structure. *Transactions of the Japan Society of Mechanical Engineers, A* 79 (800): 489–496.

Nakamura, S., Miyaoka, R., and Nakagawa, S. (2013). Effect of cooling conditions on residual warp deformation of laminates consisting of epoxy resin and various substrates. *Journal of the Institute of Electronics Packaging* 16 (3).

Nakamura, S., Suehiro, K., and Katsuyama, Y. (2013). Experimental and Theoretical Investigation of Warp Deformation Behavior of Viscoelastic Three-layer Laminate Under Thermal Load, Surface Mount Technology, (Korea) May.

Nakamura, S., Miyaoka, R., and Kondo, A. (2013). Effects of Cooling Conditions on the Amount of Residual Warpage of Laminates Consisting of Epoxy Resin and Various Substrates, Surface Mount Technology, (Korea) September, 2013.

Nakamura and Katsuyama (2016). Prediction of warpage deformation in thermal load process of four-layer laminate containing resin. *Transactions of the Japan Society of Mechanical Engineers, A* 82 (837).

Things Related to Photoviscoelasticity

Read, W.T. (1950). Stress analysis for compressible viscoelastic materials. *Journal of Applied Physics* 21.

Dally, J.W., Riley, W.F. et al. (1959). A photoelastic approach transient stress problem employing low-modulus materials. *Journal of Applied Mechanics* 26.

Williams, M.L. (1962). *On the Stress and Strain Relation with Dynamic Birefringence*, GALCIT SM 62-5. California Institute of Technology.

Hayashi, T. and Gozu, J. (1979). Method for determining characteristic coefficients of linear photoviscoelastic materials. *Transactions of the Japan Society of Mechanical Engineers* 45 (263).

Hori (1973). *Viscoelasticity*. Baifukan.

Isshiki (1976). *Heat Transfer Engineering*. Morikita Publishing.

Index

b

ball grid array (BGA), OMPAC type 52
ball grid array-chip scale package (BGA-CSP) 2, 3
Boltzmann superposition principle 40–41

c

chip scale package (CSP) 51
chip scale package-μball grid array (CSP-μBGA)
  comparison of residual curvature 76, 77
  cross section of 69–70
  elastomer structure 69, 70
  material and structure optimization design 73–77
  material properties 70–73
  modeled structure 69, 70
  residual stress distribution 75–76
  thermal stress 73–74
  VESAP analysis of 69–77
  warp deformation behavior 76
composite materials 12
cooling rate, effect on warpage behavior
  experimental method 159
  experimental values 164–166
  final warp deformation 166, 167
  heat transfer coefficient 163–164
  laminated body shape 159, 160
  properties of constituent materials 160
  relaxation elastic modulus 160, 161
  residual warp deformation 166, 167
  thermal expansion coefficient 160, 161
  time–temperature shift factor 162
  two-layer laminate 168–169
  VESAP analysis 160–163
creep compliance 36
creep phenomenon 30–31, 34, 36–37
curing shrinkage 119
  concept of 120
  examination procedure 121
  VESAP analysis 122–123
  warp deformation 128–132
curing temperature 148–149, 151–153, 156
curing warpage 123–128

d

deformation 18
dual in-line package (DIP) 2, 3, 80
dynamic viscoelasticity 43–48, 134–137

Polymeric Materials for Electronic Packaging, First Edition. Shozo Nakamura.

Published 2023 by John Wiley & Sons, Inc.

e

elastic failure criterion 19
elasticity 15
  energy 30
  entropy 30
  FEM analysis 16–17
  governing equations 18–19
  stress and strain 15–16, 28, 32
  and viscosity 28
elastic strain energy theory 21
elastomer 3, 5
engineering plastics 11
epoxy resin 10, 11, 27, 56–57, 70, 119
  curing characteristics of 119, 126
  curing temperature of 152–153
  heat generation temperature 129, 153–156
  loss tangent of 142
  master curve of 72
  shift factor of 143
  storage modulus of 142
  thickness of 129–130, 153–156
  time–temperature shift factor of 72, 73
  viscoelasticity properties for 126
  warpage behavior of
    four-layer laminate 108–117
    three-layer laminate 104–108
    two-layer laminate 92–104
epoxy resin/FR-4 substrate 92–100
epoxy resin/steel 101–104, 140

f

fiber-reinforced plastics filled with carbon fibers (CFRP) 10, 12
fiber-reinforced plastics filled with glass fibers (GFRP) 10, 12
finite element method (FEM) 6, 16–17
force and moment balance equation 18
four-layer laminates 108–117

g

glass transition temperature 136–137

h

heat transfer coefficient 96–97, 163–164
Hooke's law 12–13, 28, 32

l

large-scale integrated (LSI) package
  changes in volume of 2, 3
  cross-section of 3, 4
  design issues 49–53
  future 53
  OMPAC type 51–52
  outline transition of 2
  product design of large-scale 27
  types of 3–4
lateral strain 16
law of elastic breakage
  elastic strain energy theory 21
  maximum main strain theory 20
  maximum principal stress theory 20
  maximum shear stress theory 20
  shear strain energy theory 20–21
  yield conditions 19
longitudinal strain 16
loss modulus 136
loss tangent 136, 142

m

master curve 38–40, 72, 95, 102, 106, 110, 122, 142, 161
maximum main strain theory 20
maximum principal stress theory 20
Maxwell model 31–37
μBGA 5–6 *see also* Chip scale package-μball grid array (CSP-μBGA)
Mises' conditional expression 20

n

naphtha 9
Newton's law 13, 28, 32

o

over molded pad array carrier (OMPAC) package structure 51–52

p
panel level package (PLP) 6
pin grid array (PGA) 2, 3
plane strain 21
plane stress 21
plastics 9
Poisson's ratio 16, 127
polyethylene (PE) 9, 11
polymer materials
  definition of 9
  properties of 12–13
  types and classification of 10–12
polypropylene 11
printed circuit boards 1, 3, 27, 49, 92–100
Prony series 39–40, 61, 95

q
quad flat package (QFP) 2, 3
quenching 61–62, 64–65

r
relaxation modulus 36, 122, 142
residual stress distribution
  of CSP-μBGA 75–76
  of three-layer laminate 88, 89
  of two-layer laminate 65
resins 40 *see also* Epoxy resin
  classification and types of 10
  properties of 12, 27
  thermoplastic 11, 13
  thermosetting 11–12

s
semiconductor materials 5–7
semiconductor package 1–4, 27
shear strain energy theory 20–21
small outline J-leaded package (SOJ) 2, 3
solid body, mechanical behavior of 29–30
spring 28
storage modulus 135, 142
strain
  elastic 15–16, 28, 32
  plane 21
  viscoelastic 29–32
strain–displacement relational expression 18–19
stress *see also* Thermal stress
  difference from strength of materials 23–24
  elastic 15–16, 28, 32
  evaluations, with defects 23–26
  plane 21
  viscoelastic 29–32
stress concentration 24
stress intensity factor 25–26
stress relaxation function 36
stress relaxation phenomenon 31, 32, 36–37
stress–strain relational expression 19
superposition principle 40–41

t
thermal degradation 133–134
  change in viscoelastic properties 140, 142–143
  dynamic viscoelasticity test 134–137
  fracture surface after thermal aging 137–139
  specimen and heat load conditions 134, 135
  static tensile properties 137, 138
  two-layer laminate 139–146
  warpage deformation behaviors 140, 143–146
thermal stress
  change in 6–7
  of CSP-μBGA 73–74
  generation mechanism 41
  of three-layer laminate 87
thermoplastic resin 10, 11, 13
thermosetting resin 10–12

thin small outline package (TSOP) 2, 3
three-layer laminate
steel/epoxy resin/printed board 104–108
VESAP analysis of 79
curvature change 87
material properties 85
model 80–81
relaxation modulus of resins 85
residual stress distribution 88, 89
structure of LSI package 79–80
temperature change 88, 89
temperature distribution 81–84
thermal stress 87
time–temperature shift factor 85, 86
warpage deformation 81–90
time–temperature conversion rule 38–39
time–temperature shift factor 39, 85, 86, 123, 162, 175
transfer molding method 11–12
Tresca's conditional expression 20
two-layer laminate
VESAP analysis of 55
cooling experiment 61–62
laminate of encapsulating resin/Fe–Ni alloy (42 alloy) 57, 65–67
laminate of epoxy resin/Al alloy 56–57, 63–65
method 61
residual stress distribution 65
results and experimental values 63–67
validation procedure 55–56
warpage deformation 65–67
warpage behavior of
effect of cooling rate 168–169
effect of thermal degradation 139–146
epoxy resin/FR-4 substrate 92–100
epoxy resin/steel 101–104

V
viscoelasticity 5, 29, 32
Boltzmann superposition principle 40–41
creep phenomenon 30–31, 34, 36–37
dynamic 43
master curve 38–40
Maxwell model 31–37
measurement
complex modulus 45, 47
frequency dispersion and absorption 46–47
mechanical model 44–46
method 43–44
temperature *vs.* viscoelastic properties 48
vector diagram 46
mechanical characteristics of 30
stress and strain 29–32
stress relaxation phenomenon 31, 32, 36–37
thermal stress 41
time–temperature conversion rule 38–39
Voigt model 31–37
visco-elastic stress analysis program (VESAP) analysis 41, 50, 55
analysis time of 172
basic formulas 172–175
of CSP-μBGA 69–77
curing shrinkage 122–123
development needs and concepts 171–172
effect of cooling rate on warpage behavior 160–163
method and conceptual diagram 175
overview of 178
of three-layer laminate 79
curvature change 87
material properties 85
model 80–81

relaxation modulus of resins 85
residual stress distribution 88, 89
structure of LSI package 79–80
temperature change 88, 89
temperature distribution 81–84
thermal stress 87
time–temperature shift factor 85, 86
warpage deformation 81–90
time-varying temperature distribution 175–177
of two-layer laminate
cooling experiment 61–62
laminate of encapsulating resin/Fe–Ni alloy (42 alloy) 57, 65–67
laminate of epoxy resin/Al alloy 56–57, 63–65
method 61
residual stress distribution 65
results and experimental values 63–67
validation procedure 55–56
warpage deformation 65–67
viscosity 28–29
viscosity coefficient 29
Voigt model 31–37

w

wafer level package (WLP) 6
warpage deformation 6
of CSP-μBGA 76
due to hardening and heat 130–132
effect of cooling rate 159–169
evaluation method for 147–157
experiment 128–130
during heating and cooling process 91–92
four-layer laminate 108–117
three-layer laminate 104–108
two-layer laminate 92–104
thermal degradation effect 140, 143–146
three-layer laminate 81–90
two-layer laminate 65–67

y

yield criterion 19
Young's modulus 127